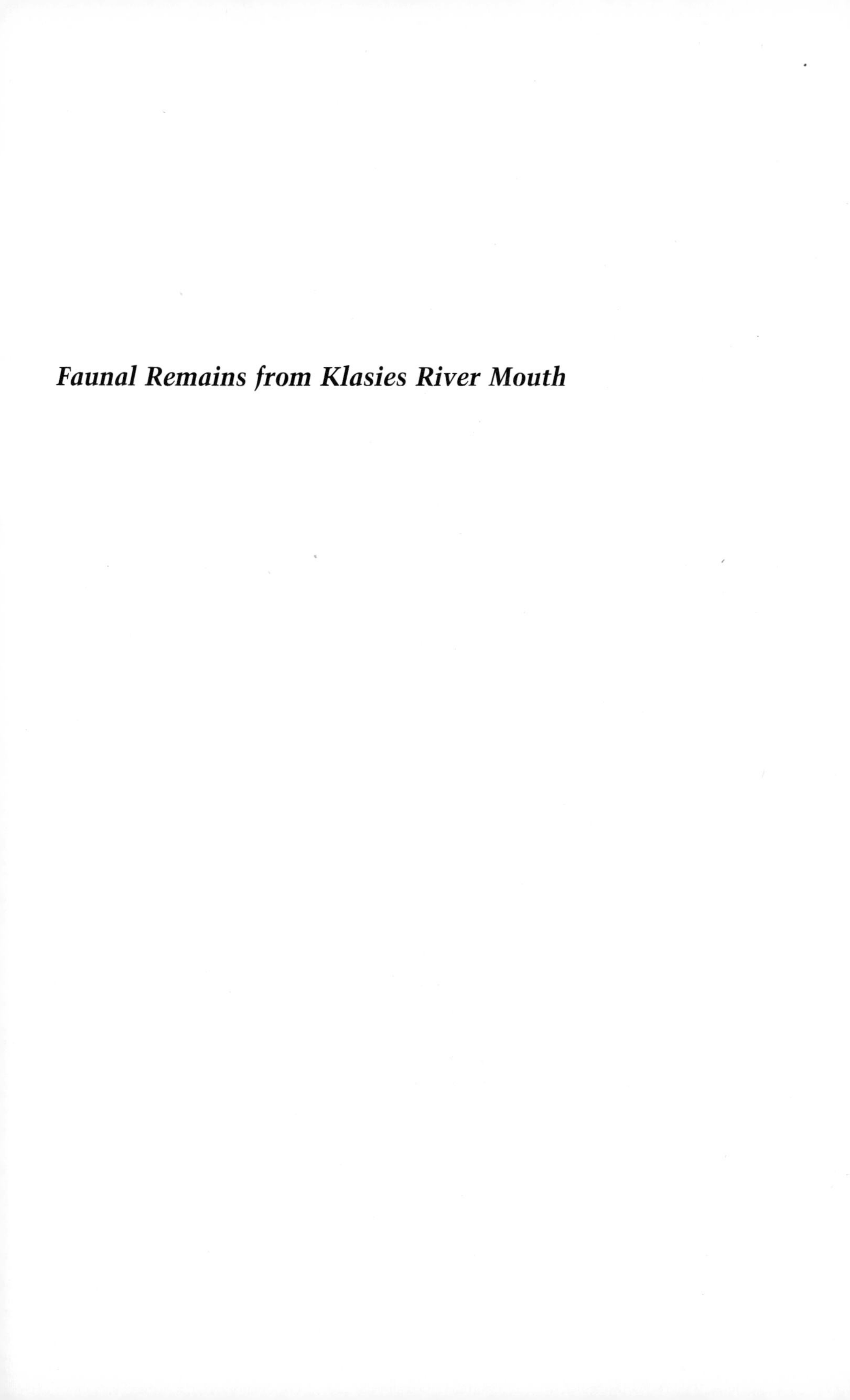

Faunal Remains from Klasies River Mouth

This is a volume in

Studies in Archaeology

A complete list of titles in this series appears at the end of this volume.

Faunal Remains from Klasies River Mouth

LEWIS R. BINFORD

Department of Anthropology
University of New Mexico
Albuquerque, New Mexico

1984

ACADEMIC PRESS, INC.

HARCOURT BRACE JOVANOVICH, PUBLISHERS

Orlando San Diego San Francisco New York London
Toronto Montreal Sydney Tokyo São Paulo

ACADEMIC PRESS, INC.
Orlando, Florida 32887

United Kingdom Edition published by
ACADEMIC PRESS, INC. (LONDON) LTD.
24/28 Oval Road, London NW1 7DX

Library of Congress Cataloging in Publication Data

Binford, Lewis Roberts, Date
Faunal remains from Klasies River mouth.

(Studies in archaeology)
Bibliography: p.
Includes index.
1. Paleolithic period, Lower--South Africa--Kaapsedrifrivier Valley. 2. Animal remains (Archaeology)--South Africa--Kaapsedrifrivier Valley. 3. Human evolution. 4. Kaapsedrifrivier Valley (South Africa)--Antiquities. 5. South Africa--Antiquities. I. Title. II. Series
GN772.42.S68B56 1984 573.2 83-15909
ISBN 0-12-100070-2 (alk. paper)

PRINTED IN THE UNITED STATES OF AMERICA

84 85 86 87 9 8 7 6 5 4 3 2 1

Contents

List of Figures

List of Tables

Preface

This book started as an article. That it grew is, in fact, part of the excitement of this piece of research. All the observations on the Klasies fauna were made in Capetown and were recorded in my notebooks. No summary tabulation or syntheses of data were attempted while I was still in Africa, nor was any attempt made at such syntheses on my immediate return. When I decided to write up the Klasies data, I started to assemble all the observations anatomically, part by part. I had no idea that the patterning reported in Chapter 4 was as robust as it turned out to be. In fact the relationships between anatomical part, animal gnawing, butchering-mark frequencies, and all the facts that have proved so interesting when synthesized for the Klasies fauna came as surprises as the descriptive work progressed. As each newly recognized pattern emerged, I frequently found myself pursuing literature on subjects not anticipated as relevant when the work began. This book, then, is the result of a productive feedback between the emergent patterning of the data and my thoughts about the implications of the newly recognized patterns for our ideas about Middle Stone Age hominid behavior.

The result is a book that develops an argument about the behavior of Middle Stone Age hominids; these arguments have strong impact on many contemporary ideas about the course of human evolution. The impact is both methodological and substantive. The suggestions that are presented regarding the factors that shaped our own humanity are new and certain to stimulate controversy. If it is followed by research instead of by misleading rhetoric, the controversy that is almost certain to appear could take us a long way toward a more accurate understanding of our evolutionary past. I am excited about this book; some of the proposals and conclusions may seem to reach too far "beyond the data." The reader may judge and develop these ideas as he or she wishes. I hold the view that conservatism will never lead us to the productive recognition of our ignorance nor to the pursuit of the research needed to reduce it. The arguments presented here are an effort to move toward both of the above goals.

Acknowledgments

The work reported here could quite literally never have been done if it were not for John Parkington and his kind invitation to me to visit and work at the University of Cape Town, South Africa. John was instrumental in getting me to South Africa; many other kind colleagues enhanced my visit and made it one of the most memorable experiences of my life. To John and all the staff and students at the University of Cape Town, I am most grateful.

Specifically, the work with the fauna reported here was made possible through the cooperation of Richard Klein of the Department of Anthropology, University of Chicago. Richard had previously studied the Klasies fauna and made the arrangements for me to carry out the observations upon which this study is based. Richard helped get me started and was always helpful when I encountered problems during the work. He introduced me to Q. B. Hendey, who also lent his support to my work. In general the staff of the South African Museum greatly contributed to the success of my work. I owe a very special thanks to Mr. E. B. Shaw, who helped me during my research at the South African Museum in so many ways.

Most of my observations on the fauna were done alone; however, on several occasions, students from the University of Cape Town went with me to the South African Museum and actually worked on the bones. For this assistance I am most grateful. During the last days of my stay in South Africa, John Lanham worked with me nearly all the time and took all the documentary photographs of the Klasies fauna, many of which appear in this book. I am most grateful for John's skill and dedication.

Back in Albuquerque I have been assisted by many of my students: Steve Kuhn, Mary Stiner, and Erik Ingbar read and commented on sections of the manuscript. Neale Draper and Del Draper both have helped with the manuscript and Del did all the typing. Charles Carrillo did the drawings that appear here as Figures 3.13 and 5.6. As usual, I have had the support of my department, and Jerry Sabloff in particular has been most encouraging.

In May 1982 I had the opportunity to lecture at Berkeley and was hosted by Glynn Isaac, F. C. Howell, and William Woodcock. It was largely the result of discussions with Tim White, Owen Lovejoy, and Glynn Isaac that inspired me to get on with my analysis of the Klasies data, which had been sitting since my return from South Africa. It was in the context of the stimulation at Berkeley that I saw more clearly the importance of the data to be reported here. Clearly many of these men will not welcome this analysis with open arms; nevertheless, it was in the context of their ideas that the importance of these observations seemed to rest.

Finally, a first draft of this manuscript was circulated to Richard Klein, Don Grayson, Tim White, J. Desmond Clark, and F. C. Howell. All responded with helpful and astute criticism. As a result of their suggestions and criticisms, I substantially rewrote Chapters 1–3, as well as 5. Most of these readers will see that I took their suggestions to heart and I have tried to answer their questions and to elaborate the points that they considered underdeveloped in the first draft. Perhaps the only reader who may be a little put out by my failure to heed some of his observations is Tim White. Tim has some strong opinions about the history of ideas in African studies. I am in agreement with most of his views. In spite of my agreement, however, I believe that the logic of my presentation in Chapter 1 has merit, and even though it is true that Louis Leakey basically always looked for earlier and earlier forms of true men, his behavior at the Darwin centennial and shortly after the discovery of the Zinj floor has not been previously summarized. I think it was fascinating in spite of the fact that the outcome (*Homo habilis*) was consistent with his earlier biases. In any event, there are plenty of materials for one interested in the intellectual history of Early Man studies; my discussion is not meant to be exhaustive.

To these persons and all who have sought to know the past, and who by their work have motivated others to get on with science, I express my sincere thanks and register my admiration and respect.

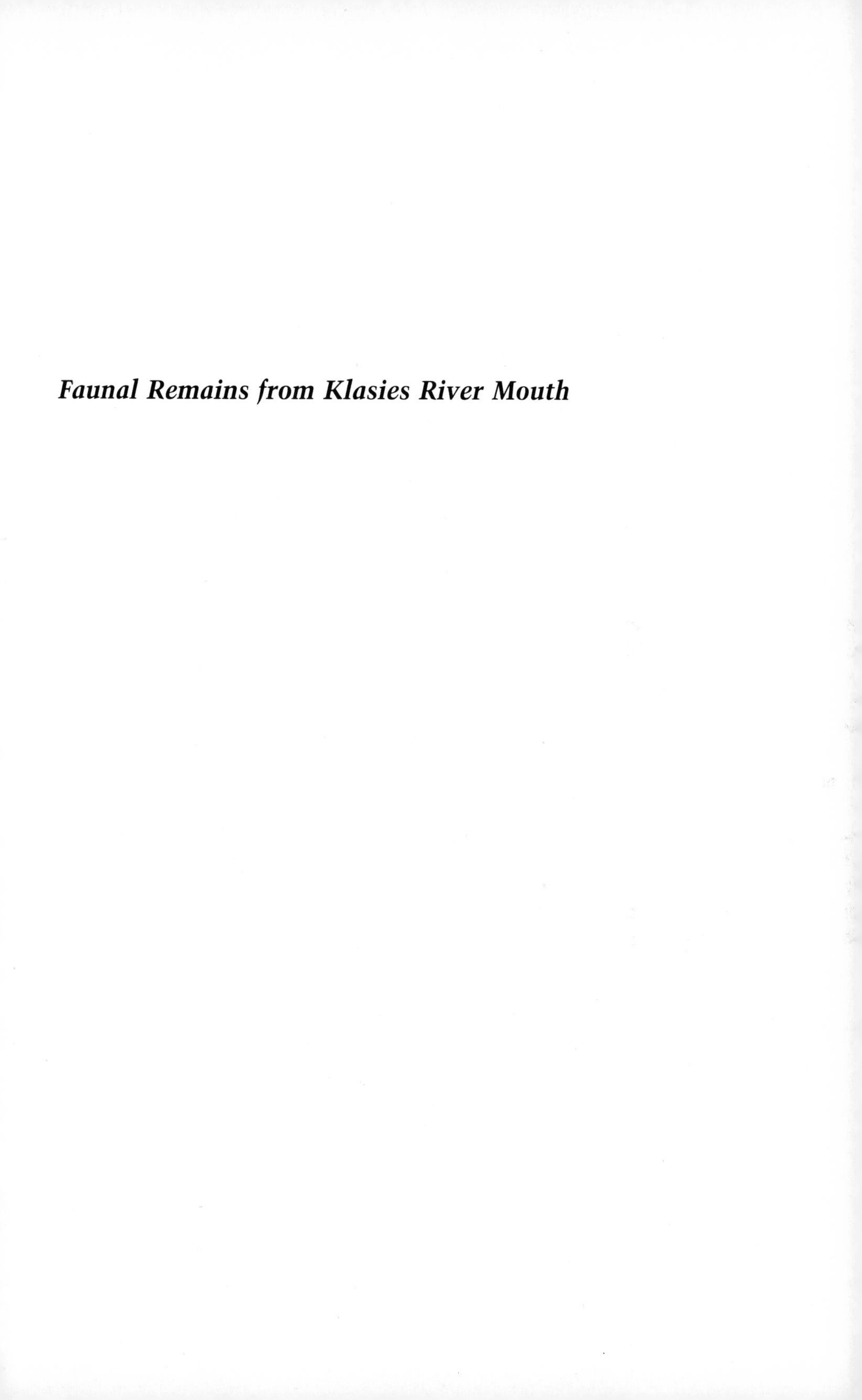

Faunal Remains from Klasies River Mouth

CHAPTER **1**

Problem, Approaches, and the Process of Learning

The Problem

Early speculations on the evolutionary transformation from "ape" to "human" generally took some form of functional argument. For instance, Carveth Read (1920), arguing before australopithecine fossils were known, suggested that precultural man was a pack-hunting ape subsisting upon moderate- and large-size mammals. He speculated on how tool-using members of this pack would have an advantage, as would those individuals psychologically prone to cooperative behavior. These individuals would develop leadership and its complement, a disciplined group of followers (Figure 1.1).

Many subsequent writers, considering the problem after more knowledge of the australopithecines became available, argued that bipedal locomotion gave the hunter a competitive advantage that preceded the evolutionary appearance of the typical "human" brain. Most who pondered the course of hominid evolution held that the australopithicines represented a transitional period between an earlier prehuman form of brain and the far more complex mentality that enabled humans to develop the culture that removed them from the animal world. Almost without exception, earlier speculation on this biocultural transformation gave an important conditioning role to hunting, which moved our prehuman ancestors into the predatory niche. (For a good example of this approach, see Etkin 1954.) Serious

Figure 1.1 A pack-hunting model of early hominid life.

speculation on the early hominids' survival tactics in obtaining food frequently touched upon the possible role of scavenging. For instance, in their provocative and important essay, Bartholomew and Birdsell (1953) offered the following suggestions:

> It is difficult, perhaps impossible, to determine whether or not the remains of the large giraffids and bovids reported from the bone breccias (Dart 1940) represent kills by australopithecines or their scavenging from the kills of the larger cats. Since few meat eaters are loath to scavenge, and the implementation which would allow the australopithecines to kill such large animals is not apparent, we suggest that scavenging from the kills of the larger carnivores may have been systematically carried out. (Bartholomew and Birdsell 1953:490)

This suggestion—favoring a "scavenging" or, more appropriately, "scrounging" phase over the gradual shift from a nonpredatory niche to the behavior of the predatory hunter—appealed to those who viewed evolution as the playing out of a gradualist's belief that one must crawl before one can walk.

A gradualist position was adopted by Louis Leakey at the time of the discovery of the "Zinjanthropus" fossil (now known as *Australopithecus boisei*).

> On July 17, 1959, my wife, working with me at Olduvai Gorge in Tanganyika Territory, found a fragment of fossil human skull on the slopes of the gorge at site FLK I . . . When excavations were carried out, a nearly complete skull of a hominid and also a tibia were found. These were lying on a living floor upon an ancient camp site, in association with nine stone tools of the Oldowan culture and 176 waste flakes, which had resulted from the manufacture of the tools on the spot. . . . Associated with these tools and flakes were the fossilized bones of *many small*

creatures, such as rats, mice, frogs, lizards, birds, fish, a snake and a tortoise, plus the bones of some juvenile pigs and antelope and a juvenile giant ostrich. (Leakey 1960:24; emphasis added; © 1960 by the University of Chicago)

The Darwin centennial—the hundredth anniversary of the publication (1859) of *The Origin of Species*—was poised on the edge of a major intellectual turning point in the arguments relating to the evolutionary history of man. The older views were dominant; yet the discovery of the "Zinj" floor by Mary and Louis Leakey had recently been made. The statements by Leakey represent the accommodation of the new facts to the earlier views of man (see particularly Leakey 1960). Similarly, the summaries by the other major researchers document nicely the ideas that provided the intellectual background for a number of "new" arguments made by Louis Leakey shortly after the Darwin centennial.

Leakey's post-Zinj claims have intellectually framed early man research since the decade beginning in 1960. At the time of the dramatic discovery of the Olduvai material, most students of human evolution had rejected Raymond Dart's (1948) claims that the australopithecines were predatory hunters,—just as they rejected, two decades earlier, his claim that the australopithecines were hominids.

At the same time, the majority accepted the position, increasingly solidified by S. Washburn (1959), that hunting was the crucial behavioral context in which humanization occurred, leading to that culture-bearing creature we know as modern man, or *Homo sapiens sapiens.* Critical to the development of events not yet discussed was the linkage made in the Washburn model between bipedal locomotion, which made possible the shift to hunting, and hunting, seen as the selection context that favored increased brain size and the appearance of such basic human social characteristics as the division of labor, food sharing between adult males and females, and a growing dependence upon information transmitted through learning. This position was summarized by Washburn and Howell (1960:49) at the Darwin centennial meeting in Chicago as follows:

> It would now appear, however, that the large size of the brain of certain hominids was a relatively late development and that the brain evolved due to new selection pressures after bipedalism and consequent upon the use of tools. The tool-using, ground-living, hunting way of life created the large human brain rather than a large-brained man discovering certain new ways of life. The authors believe this conclusion is the most important result of the recent fossil hominid discoveries and is one which carries far-reaching implications for the interpretation of human behavior and its origin. (© 1960 by the University of Chicago)

During the 1960s, it was commonly held that early hominids' consumption of small animals, supplemented by scavenging larger carcasses, represented a gradual transition between the more apelike nonpredatory niche and the hunting behavior that was thought to be the molding condition of our humanity.

> Australopithecus was living on small reptiles, birds, and small mammals (such as rodents). As well as presumably on roots and fruits. . . . Only in Middle Pleistocene do we find evidence of a major change in early man's adaptation to plains living, and this change involved cooperative hunting—a change in food-getting behavior of central importance to the story of human evolution. (Campbell 1966:201–202; copyright © 1966 by Bernard C. Campbell. Reprinted with permission from *Human Evolution* [New York: Aldine Publishing Company.])

Most of these points were consistent with the then-persuasive view of S. Washburn (Washburn and Lancaster 1968) regarding the importance of hunting as the behavioral context for selection pressures leading to our human condition (see Howell 1965, particularly pp. 64–65). It was within the perceived sequence of hunting first, followed by a development of the brain, linked with the idea that the australopithecines were the transitional stage in which hunting was foreshadowed by the casual eating of small animals and the occasional scavenging of meat from carnivore kills, that Louis Leakey expanded his investigations at the now-famous site of the Zinjanthropus discovery—FLK-22.

During the early years of the 1960s, a major shift took place in the thinking regarding our hominid evolutionary background. The data from Mary Leakey's detailed excavation at the Zinj floor seemed unequivocal in pointing to a more substantial role for hunting in the subsistence base of the hominid responsible for the FLK-22 "living floor" than would have been previously anticipated. Animal bones from a variety of large- to medium-size bovids were scattered in considerable density on the premises; we heard no more of baby pigs and birds' eggs.

If one accepted the association of stone tools and animal bones as indicators of a living floor, then two inconsistencies seemed evident: (1) Zinjanthropus was relatively small-brained, yet his success at hunting seemed evident; (2) if one believed that increases in brain size were a consequence of hunting behavior, then his brain should have been larger. In any event, Zinjanthropus was not a "transitional" form. In short, there appeared to be an inconsistency between the dominant belief about the behavioral contexts of encephalization and the archaeologically indicated behavior.

What followed was a complicated series of arguments mounted by Louis Leakey (1965), which questioned almost all aspects of the situation *except* his belief in what was thought to be the self-evident meaning of the associated bones and the stone tools—namely, that these were living floors or home bases referable to an imagined life-style that differed little from a watered-down picture of modern hunter–gatherer peoples: "it seemed natural to treat these accumulations of artifacts and faunal remains as being 'fossil home sites'" (Isaac 1983:1).

The apparent incompatibility between the evidence for successful hunting and the size of the brain of Zinjanthropus was resolved by the Leakeys' argument that an actual ancestor of modern man had lived contemporaneously at Olduvai. These larger-brained creatures, designated *Homo habilis* ("Handy Man"), were considered the successful toolmaking hunters who witnessed the extinction of Zinjanthropus. This viewpoint saved the original theoretical scenario, in which hunting and food sharing with their attendant socialization provided the context of our humanization. The theory had been correct; only the initial view that the australopithecines were transitional had been wrong. They were, on reconsideration, seen as a collateral line existing alongside the more adept habilines who, due to larger brain size, with its implied link to humanlike behavior mented the generic status of *Homo*. Certainly the confidence in the theory was greatly enhanced later, when comparatively large-brained habiline specimens, such as the famous 1470 skull, were found (R. Leakey 1973). This seeming support for the theory lent more credence to the search by archaeologists for earlier humanlike behaviors. In fact, their existence even appeared necessary and plausible.

Such, then, has been the intellectual context for most of the recent work conducted on sites of the Plio–Pleistocene time boundary. Glynn Isaac, summarized the situation this way:

> If we accept the working hypothesis that early tool-making hominids sometimes transported food to campsites, or temporary home bases, where they also made and discarded stone tools, then we face the challenge of developing models of probable socioeconomic systems that incorporate such behavior. . . . As we see it the pivotal ingredient is *active food-sharing,* with some food being transported to a shifting but well-identified spatial focus that can be termed a *home base.* In the versions of this model practiced by living peoples, *division of labor* between *hunter–scavenger–fishers,* who are preferentially female, is virtually universal. The system cannot operate without some simple *equipment and tools,* namely containers for carrying food and knives to cut up carcasses. The whole complex is interconnected with an evolutionary change in anatomical locomotor arrangements, namely *bipedalism,* which facilitates *carrying* things about. A social system involving exchange of energy in the form of transported food puts a premium on the ability to exchange information and make arrangements regarding future movements of group members (Isaac 1976a–d, 1978). It also increases the importance of regulating social relations among individuals. All these influences might be expected to favor the evolutionary development of an effective communication system, such as protolanguage. (Isaac and Crader 1981:90–91; original emphasis; © 1981, Columbia University Press. Reprinted by permission.)

One can see that these are reasoned arguments of Pleistocene home bases. They rationally consider how different behaviors and properties are mutually dependent upon one another and how they mutually condition the appearance of behavior syndromes. In the scenario given above, the only two characteristics that really require testing historically are bipedal locomotion and the idea that hominids occupied home bases. Given these "facts," all the other behaviors are seen as flowing logically from these preconditions.

What is the role of the archaeological record in testing such scenarios of functional reconstruction? The answer is really very simple: it is to warrant the belief in one or more of the pivotal conditions from which the remainder of the reconstructive argument proceeds. Only if the early sites can be defended as home bases and the animal bones found therein justified as the products of hunting, or at least the very successful scavenging of large meaty parts, can the theory of evolutionary cause and effect be supported.

I have argued (Binford 1968, 1981, 1983a,b) that the manner of generating linkages between concept and experience, theory and observation, dynamics and statics, is central to archaeology; indeed, it is the most critical form of research in which archaeologists can be engaged. I am of the opinion that those archaeologists who sought to use the archaeological record as a source for testing the consensus theory treating the emergence of a "thinking" man, did so generally in the absence of any well-thought-out attention to this problem of linkage—the problem of middle-range justification for the meanings that were attached to archaeological observations.

Unquestionably, Glynn Isaac is the leading archaeologist in this research area. He is also the one most closely associated with testing the hunting, food-sharing, home-base explanation of humanization. But how has Isaac proceeded to test this theory?

Before responding to that question, I need to generalize—and philosophize. Theories can be thought of as trial statements about the way dynamics could be organized. Thus, almost all theoretical terms take their meanings from the other ideas and concepts within the argument having reference to causal dynamics. When archaeologists appeal to the empirical world of the archaeological record in order to evaluate a theory, they must face quite directly the problem that theories of history are generally about dynamics; yet their empirical world—the archaeological record—consists of static, structured arrangements of matter. Most theoretical terms are primarily defined in terms of other concepts, not in terms of empirical properties. All theories address organized interactions and/or mutually conditioning situations conceived to cause or shape the values of several variables. Variables, as important elements in theories, almost always refer to dynamic phenomena or dynamic system-state properties, which can be thought of as properties abstracted from our experiences with the dynamic histories of organized systems.

If the archaeologist seeks to test a theory by appealing to archaeological facts, how is this going to be achieved? Theories describe dynamics, conditions of organized systems, whereas archaeological facts are all static properties of matter. Minimally, the archaeologist must operationalize definitions for the theoretical terms or variables in the theory, in the context properties that could be experienced in archaeological statics. In short, he or she must argue that there is some justification for translating theoretical terms—having reference to dynamics and particular intellectual-conceptual frameworks (theories)—into expectations regarding particular arrangements of static archaeological matter.

Now, how has Glynn Isaac sought to solve this problem? His strategy seems to be this: If we can identify these home bases in the archaeological record, then by implication we have evaluated the arguments regarding the dynamic and causal linkages between the behaviors believed to stand behind such phenomena as home bases, and hence have provided credence and support for the theory. Isaac initially offered an operational definition of *home bases* as those archaeological sites that exhibit high densities of both animal bones and stone tools. Sites where stone tools were common but bones rare were identified as *quarry* or *workshop* sites, whereas high bone-density, low tool-frequency sites were *kill* or *butchery* sites. Those with low densities of both bone and stone were transitory camps (see Figure 1.2).

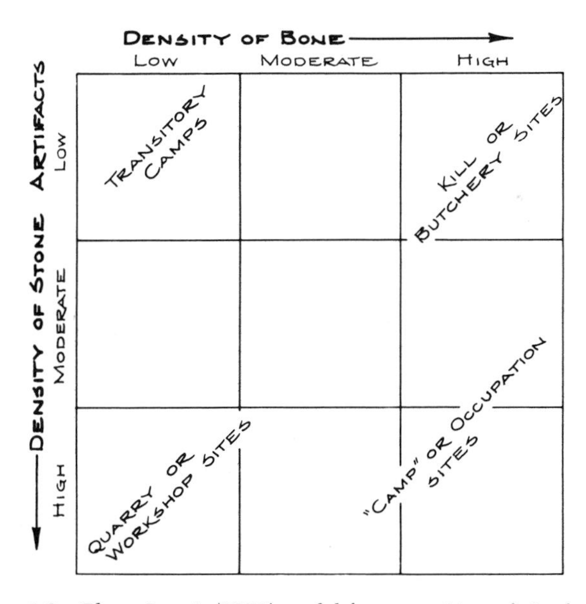

Figure 1.2 Glynn Isaac's (1971) model for recognition of site function.

Here we see a very interesting approach to inference. Specifically, the middle-range argument is that if archaeological sites exist (defined by concentrations of stone tools) in association with dense patches of animal bone, then the home base is identified and, by implication, the entire theory of humanization is thought to be strengthened. Interestingly, we already know that such sites existed. In addition, the demonstration that some sites were not home bases in no way disproves the existence of home bases themselves, nor does it diminish the plausibility of the investigator's belief in the cultural efficacy of food sharing, communication, and hunting as contexts of selection for the gradual transformation into the human state. We have here a middle-range procedure that is a complete tautology. The archaeological record cannot talk back; it can never, regardless of what is found empirically, negatively affect the belief in the sharing hypothesis. This is not science. Minimally, *science* is the development of means whereby experience with the world can be used as an arbiter, a means to evaluate our ideas and theories. Isaac has often reiterated his "theory" and his middle-range justification for believing it, and the basic structure of the argument has remained unchanged.

I have already questioned (Binford 1977b) the accuracy of attributing *all* the properties of one of these recognized "living sites" or "home bases" to the actions of hominids. I suggested that Isaac had not treated very analytically the possibility of other active agents operating in the past so as to contribute to concentrations of matter that were being interpreted as if man was the only formative agent. Mine was an early challenge, though other scientists also had proposed the possibility that other agents might have contributed to the formation of these patches of stone tools and associated objects.

If we are really going to test theories about the causes for, or the behavioral contexts of, our evolution, we must move to the challenging task of evaluating the models—those interpretative conventions permitting us to convert archaeological observations into statements about the past. We should be able to test the degree that the criteria we use for recognizing home bases are indeed unambiguous and are justified independently of the model they are used to evaluate. For instance, the operational definition of a home base given by Glynn Isaac does not consider that any other organized conditions of hominid life could result in a relatively dense, clustered association of bones and stone tools. In short, for all of Isaac's commitment to the method of multiple working hypotheses, he has in the past worked from only one interpretative model. To him, the bone–stone association means home bases—or, as he is currently discussing it, support for the "central place foraging hypothesis" (Isaac 1983b:11).

I visualize the methodological challenge as one in which we must conduct the research necessary to diagnose quite directly the dynamic characteristics of the past. Any synthesis of what life was like, or what the nature of the hominid niche consisted of during the early time ranges, must rest with our ability to develop independent instruments for diagnosing dynamic characteristics of theoretical interest. Our picture of the past must be built up—synthesized, if you will—from the various independently justified readings, or frameworks for inference, that we might obtain from the archaeological remains, using our instruments for measurement.

Given the importance of the role of hunting in the various arguments of evolutionary functionalism and the importance of associations between stone tools and animal bones to the current conventional methods for inference used by archaeologists, I seek to refine ways to reduce ambiguity. I seek ways of reliably distinguishing hunting from scavanging as the behavioral background for animal bones utilized by man.

As we all know, to desire knowledge is not enough; we must have reliable ways of gaining new knowledge. However, the stipulations of conventional interpretative methods are suspect as a basis for inference.

Approaches to Research

My job as a scientist is simply to be productively engaged in the pursuit of knowledge. To be actively involved in a search for knowledge means that we must recognize the nature of our ignorance. We might even say that the pursuit of knowledge requires the identification of ignorance. Given such a recognition, our goal becomes the transformation of ignorance into a better-grounded—or at least less ambiguous—form of understanding; in short, the rendering of ignorance into knowledge. This is a very big order indeed. How do we go about accomplishing such a seemingly miraculous transformation?

In the present work, I am concerned with seeking knowledge about a domain of past human behavior about which most would acknowledge that we are ignorant. In addition, I am concerned with reducing the ambiguity that might surround certain types of potentially relevant observations. Finally, I am interested in using the knowledge (or clues to knowledge) newly generated to enlighten both our prior archaeological observations and our ideas about what the ancient past was like. In short, I seek one form of knowledge for the purpose of developing methods useful for making inferences from the archaeological record.

When I learned archaeology, the conventions used for assigning meaning to archaeological observations were not generally under investigation. They were taken for granted. The only domain about which archaeologists readily admitted ignorance was the past. Archaeologists acknowledged that we needed to know more about this time period or this region or that to fill out our picture of the past. We investigated the past through the archaeological record, which was understood in terms of a series of conventions. I challenged this view. I tried to suggest that there were many different conditions in the past that might well structure an archaeological record. The past reality and its archaeological remains, when subjected to interpretation by archaeologists using the conventions of the time, would be distorted and misrepresented. I argued that the conventions of the day were most likely inadequate, and certainly were clouded by ambiguity (see Binford 1963).

Later (Binford 1981), I strongly criticized many suggestions as to how we might justify inferences to the past. My main thrust was to point out the trap of adopting plausible suggestions as conventions for making inferences from archaeological observations about the nature of the past. Most of the time, the impudence of conventionalism derives from incomplete knowledge. That is, an investigator adopts some suggested connection between one set of conditions and another and then assumes that this suggestion is complete and accurate. It is assumed that the linkage between one thing (the cause) and another (the effect) is unambiguous (nothing else could equally lead to the effect as observed), or the relationship is necessary (the cause always and necessarily generates the specified result and nothing else could do so). When it is possible to demonstrate that these assumptions regarding the relationships between one condition and another are unfounded, it is always because we have gained knowledge that was not available to, or was ignored by, the earlier worker. Given new knowledge, we can see in retrospect that a spurious or ambiguous link was made between two conditions. Retrospective criticism is thus made possible by a growth in our knowledge.

Progress in science happens because curious people are willing to risk suggestions as the nature of linkages among various forms of phenomena. These hypotheses are then available for research, to be investigated as to whether the knowledge that served originally to warrant the linkage was adequate and accurate. Most of the time we find that the original premises were poorly founded (see Binford 1983c). However, such a judgment can only be rendered by virtue of bringing to bear increased knowledge and understanding relative to the original suggestions. In effect, we succeed in our pursuit of knowledge as we proceed along. This is commonly how we learn: we are prompted by a bold companion who suggests that his knowledge is adequate to warrant a proposition that links alleged causes and effects. Perhaps we are skeptical of the suggested linkage, or believe that the knowledge cited as a warrant for accepting the hypothesis is inadequate. Given such a focus of suspected ignorance, we then commence to investigate both the alleged linkages and the adequacy of the knowledge previously cited as a warrant for believing them. If we are lucky, we may make new observations and conceive of new ideas about the world that either render the original suggestion obselete, or else permit its elaboration so as to reduce any ambiguity newly recognizable in the relationship as originally proposed.

The growth of knowledge results from an inductive process. It is well established that there are no rules that ensure accurate inductive arguments. Knowledge grows largely through an interactive process of exploring the consequences of ideas for experience. At the same time, one keeps a sharp eye out for implications from experience for the ideas one is working with.

The death of an established science occurs when interpretative conventions are adopted and used as if all our methodological tools for making inferences were adequate, accurate, and completely informed. This was the state that I perceived in traditional archaeology many years ago.

My rather iconoclastic attitude eventually resulted in *Bones: Ancient Men and Modern Myths* (1981). In that book, I investigated many conventions that had arisen in archaeological practice and had served as interpretative principles for constructing a past. Specifically, I was critical of conventional wisdom from the perspective of new knowledge and insights regarding the role of animals as contributors to deposits in which hominid materials were also found. In addition, I wished to explore certain warranting arguments advanced by students of fauna as to the significance of various types of breakage and forms of inflicted marks. In many ways this was a retrospective criticism, using the perspectives of new data as well as experiences with both animal and modern human use of prey carcasses. Thus it was also study of the impact of new knowledge or observations on old interpretative suggestions and conventions. I took the study even further, to suggest how a growth in knowledge actually plays a crucial role in conditioning what we accept as a "knowledge" of the past, in those processes that operate to mold the character of the archaeological record. During the course of this retrospective study, I made a number of suggestions, such as to how to seek justifications for inferences and when we should be skeptical. I also pointed to forms of argument that have frequently proved to be inadequate in the face of new knowledge.

I have been surprised by some responses to these discussions in my book. Some readers seemingly adopted my suggestions as conventions to be used in judging the truth of arguments. But this simply cannot be done. As stated earlier, new knowledge derives from a process of inductive reasoning, and I know of *no* conventions that may be followed to ensure unambiguous and accurate conclusions to be drawn inductively. My discussions regarding procedure and the history of failures in argument were offered not as conventions for judging the accuracy of an inductive argument, but instead as guides to healthy skepticism. We must identify areas of suspected ignorance before we can rationally proceed or design research aimed at reducing this ignorance.

I am convinced that the most productive clue to areas of critical ignorance derives from our skills in justifying a healthy skepticism. I do not mean a skepticism regarding what we can know—there, we must be totally optimistic—but a skepticism as to what we *think* we know and understand. It is reasoned skepticism that leads us to the productive recognition of the nature of our ignorance. It is clues to what we do not know that provide the goals for structuring a research program aimed at reducing our ignorance. Stated another way the goal of reducing suspected ignorance provides the basis for a rational assessment of our suggestions about how to pursue knowledge. If we accept this general proposition, there are other implications of importance. One is that we must be willing to use the knowledge available to us at any point in time, since it is only relative to prior claims for knowledge and understanding that we may be skeptical. In turn, it is skepticism that directs the search for new knowledge and understanding, and hence powers our success in the pursuit of knowledge. This point cannot be overemphasized. We must be willing to ride with our knowledge of the moment, for reasoned argument from this alleged knowledge actually accomplishes two things: (1) the use of the knowledge establishes that type of knowledge as important and worth having; and (2) given the perspective provided by the larger argument as to the importance of the cited knowledge, we may better focus our potential skepticism on important areas for investigation.

To illustrate this situation, I might cite a much earlier study (Binford 1978), in which I reported on facts of economic anatomy established through the observation of three animals—two sheep and one caribou. I then indicated how such facts could be used in the development of inferential methods for giving meanings to anatomical-part frequencies observed by archaeologists. Thus my arguments using these "facts" to establish important potential roles for such facts in serving the research methodology of archaeologists.

Critics have suggested that my three animals were certainly an insufficient sample to establish *accurately* the relative economic values for anatomical parts of either caribou or sheep, much less other species. I am certainly in agreement with this criticism. A critic may then ask why I was willing to build such extensive arguments on such an admittedly poor sample. The answer is rather simple: when I started the research, I had only a vague idea as to the significance of such facts. I worked to obtain sufficient facts of the type I then judged important in exploring methods development. I believe my endeavor was quite successful, because I was able to show a rather startling potential for such facts. As a sideline to this research, I became increasingly impressed with how such seemingly narrow-focused facts, using the relative values generalized from a sample of only three animals, yielded remarkable insights when different species were studied (see, for instance, Speth 1983).

This suggestion of broader relevance for facts of economic anatomy changed my ideas as to how we should proceed in obtaining facts considered adequate for grounding the methods that I was engaged in constructing. My original plan was to perform comparative studies of separate species, developing independent scales for each species of interest. These scales would have to be standardized on a large sample of studied animals, seeking to include in the sample all individual variations in nutrition, age, sex, and subspecific "racial" conditions that might conceivably affect economic anatomy.

The apparent success of my miserable little sample of three individuals representing two species, however, strongly indicates that a different mode of research might be more appropriate. Comparative study of single individuals from different species to determine how variable species were in relative economic–anatomical properties actually might be a better study program. If the levels of differentiation are slight among species, then large samples of animals within a species might be a wasteful type of study.

I have not yet decided on what I consider to be the most appropriate way of observationally grounding those facts of relative economic anatomy that thus far have proved to be of great methodological potential. I am still in the process of gaining some perspective on the problem through the increased application of the facts obtained from my observations on three animals. Needless to say, more studies of economic anatomy are certainly called for, but we cannot really judge how to conduct such studies until we gain some appreciation for the "grain" or degree of specificity at which generalizations should be targeted. It is only through the use of "knowledge of the moment" that we gain some appreciation of how best to proceed in our search for additional knowledge.

This study, then, exemplifies the realities of ongoing research. The reader will thus find me periodically appealing to ethnographic or egographic analogy, and employing anecdotal justification for accepting some propositions as knowledge. I shall also generalize from small samples and even use poorly controlled observations as operational knowledge. These are all appeals to and a use of knowledge of the moment, which is quite variable in quality and quantity.

Knowledge of the moment is all that anyone can use at any given time in developing types of argument. The uncertain and temporal grounding of such arguments is what provides the intellectual context for focusing our health skepticism. The arguments that inductively go beyond this grounding supply the intellectual stimulation for the skeptical evaluation of our present knowledge. Similarly, the inductive extension of argument frequently leads to the recognition that we need information and knowledge in a different *form* from that which grounded the original arguments. This intellectual use of knowledge of the moment in argument provides the impetus for skeptics' productive work, as well as an equally stimulating framework for sympathetic researchers who seek to expand our present knowledge. Either way we win, because the pursuit of knowledge itself is our long-range goal. As new knowledge and understanding are generated by the intellectual framework, the paradigm that guided the growth of knowledge will almost certainly become obsolete, or at least in need of revision. Only by overstepping the secure knowledge of the moment can we inductively generate a new motivating framework and provide fresh intellectual context for the further pursuit of knowledge.

In the old concept of the process of science as defined by the strict empiricists, one observed the world and then sought more and more comprehensive "empirical laws" that would eventually fit into an accumulating body of "truth"—which could be regarded as a comprehensive and integrated statement about the nature of the natural world. But this procedure ultimately stultified the imagination, which is the best source for inductively generated views that go beyond our knowledge of the moment. Meanwhile, it demanded that we keep on making observations, presumably improving the quality of our knowledge through increasingly refined observation. After all the facts were in, their true significance might be objectively recognizable; there would be no need for our imagination.

This empirically oriented procedure is now widely recognized as both impossible and counterproductive. Nevertheless, when the risk takers among us do use our imaginations and appeal to poorly grounded knowledge to build an intellectual framework to serve our goal of knowledge growth, the critics behave as if they still believe in a strict empiricist's view of the growth of knowledge; they generally try to knock down the new argument by showing that its grounding is weak. Yes, the grounding may be weak, but what are the potential gains for pursuing the knowledge required by the argument? That is the vital issue.

What I am saying here is that a healthy skepticism that questions prevailing theories is not an attitude of rejection. It is not a posture of falsification that intends to show that the facts cited in an inductive argument are insufficient to warrant the conclusions drawn—which is always the case for all forms of inductive argument. No, a healthy skepticism is a probing and constructive attitude that seeks to identify the character of our ignorance as it emerges in forms of inductive argument. We then seek to conduct research that will increase our knowledge about that very area that our skepticism has identified as perhaps inadequate or ambiguous because of some argument judged important at the time. Using such an open-minded process, a general approach to research could never fall into the nonproductive trap of sterile conventionalism that has, I fear, dominated much of the history of the science of archaeology.

The Research Tactics: Where Do We Seek Insights?

This volume is about research that seeks to develop ways of evaluating the relative roles of scavenging versus hunting in the subsistence tactics of ancient hominids. How do we do this? Ideally, I would like to go out and study a group of people who are obtaining a large proportion of their diet by scavenging. In that situation I could quite directly study the relationships between the dynamics and the static by-products remaining from various scavenging tactics. Unfortunately, I know of no opportunities for doing this. I face a situation quite common for archaeologists: I cannot gain a firsthand knowledge of many behavioral and dynamic conditions that characterized the human past by studying contemporary homologies or analogies. I must fall back, then, on a different approach. I must use what knowledge I have to tease out new knowledge and understanding.

I have previously suggested that hominid scavengers might well be expected to exploit heavily the marrow bones and perhaps the heads remaining on the sites of ravaged carcasses (see Binford 1981:266, Columns 11 and 12). I start here by seeking out an archaeological case characterized by the properties I suspect as indicative of scavenging. If such a case can be found, then I can study the fauna in detail, searching for patterning previously unsuspected. Such patterning may be in such properties as breakage, inflicted marks, and evidence of animal gnawing. Obviously, I do not know what scavenging looks like when manifested archaeologically. I do, however, have considerable knowledge of the conditions that promote different types of bone breakage, kinds and placements of inflicted marks, and forms of animal gnawing. Clearly, we must be willing to use the available knowledge, no matter how limited it may be; otherwise we will never see new things or ask new questions. Thus I can pursue a kind of alternation or interactive strategy in which I take some knowledge to help in isolating a provocative case. I might then study the provocative case in terms of properties about which I have some additional knowledge and that I suspect might well implicate diagnostic properties of scavenging versus hunting, and so forth. In short, I could work back and forth, using my knowledge to guide my observations, and then using my observations, in a newly discerned patterning, to guide my search of contemporary species for reliable understanding. The initial task, therefore, is to find a provocative fauna—a fauna characterized by a head-and-lower-legs pattern of anatomical-part dominance.

In 1976, Richard Klein published a description of the fauna from Klasies River Mouth, South Africa. One of the interesting features of the fauna described by Klein was the differential frequencies of anatomical parts recovered from the deposits. Klein introduced an argument to "explain" the differential pattern of anatomical-part frequencies observed among animals of different body size:

> the ratio of cranial to postcranial parts increases, while the ratio of limb-bones to foot-bones decreases with the size of bovid. . . . I believe that the differences . . . reflect mainly what Perkins and Daly (1968) have called the "schlepp effect." Basically they postulated that hunters were likely to bring home smaller animals intact, but they would probably bring back only selected parts of larger animals. This is because larger animals would be butchered at the place of the kill and the less useful parts would be left there. In documenting the operation of the "schlepp effect" at the early holocene ("Neolithic") hunters' site of Suberde in Turkey, Perkins and Daly showed specifically that larger bovids tended to be represented disproportionately by their foot-bones versus leg-bones just as at Klasies. They postulated that the Suberde people discarded many larger bovid limb-bones at the kill sites, but brought back the feet either as handles in the skins (used as carrying containers for the meat?) or because the feet were particularly valued, perhaps as sources of sinews for sewing. It is possible that one or the other explanation for a disproportionately high representation of larger bovid feet also pertains to Klasies. (Klein 1976:88)

What fascinated me about these interpretative arguments is that my experiences with hunting peoples of the contemporary world (Binford 1978, particularly pp. 75–90) documented a pattern of body-part abandonment that was directly opposite to that discussed by Perkins and Daly and adopted by Klein for the interpretation of his body-part patterning at Klasies River (see also Kehoe 1967:107; T. E. White 1954:256).

Among the Nunamiut Eskimo, the general condition regarding the differential abandonment of body parts is that parts of low utility are most often abandoned at kill locations, while parts of increased utility are transported to living sites. When there are transport problems, the most commonly abandoned body parts are the heads and lower legs. That is, when there is a large quantity of meat available (as with a large-body-size animal), the parts most frequently left at the kills were the heads and lower legs (see Binford 1978:76). These are the very parts that Klein notes as most commonly introduced to the site at Klasies River Mouth from large-body-size animals! This observation takes on an added interest when it is recognized that the patterning so frequently noted for modern hunters—the differential abandonment of heads and lower legs at kill sites—does not appear to be the context for the accumulation of large-animal head-and-lower-leg parts inside the rockshelters at Klasies River Mouth. Thus this site is almost certainly not a large-mammal kill site. Adding to the interest is the fact that Klein has noted the same body-size-related pattern of bias at other sites equally difficult to view as kill sites. Here we have a situation in which the interpretation flies in the face of what we know. Therefore we have a chance to learn something:

> Discovery commences with the awareness of anomaly, i.e., with the recognition that nature has somehow violated the paradigm-induced expectations that govern normal science. It then continues with a more or less extended exploration of the area of anomaly. And it closes only when the paradigm theory has been adjusted so that the anomalous has become the expected. (Kuhn 1970:52–53)

Perhaps one of the reasons I recognized the "anomaly" was that other research had led me to question the conventional wisdom regarding the role of hunting in the subsistence regime of early man (see Binford 1981, particularly p. 296). In any event, this pattern dominated by lower legs and heads is exactly the form likely to result from scavenging by hominids.

The Klasies River Mouth assemblages were an ideal case to study in the absence of actualistically controlled observation on assemblages resulting from scavenging. First, these assemblages were historically far enough removed from the Olduvai materials that no amount of advocacy one way or the other would implicate the assemblages discussed in the *Bones* book or argued about in the aftermath of its publication (see Bunn 1982; Freeman 1983; Isaac 1983a,b). Second, I could accept the evidence from Klasies River Mouth as indicative of the transport of anatomical segments of animals into a site by hominids. If scavenging could be sustained as the procurement context for these introductions, then we would have at least one model for what the much-discussed home bases should look like when scavenging was a major contribution to the diet there.

I have chosen to study the large-mammal fauna from the sites at Klasies River Mouth in hopes of learning something that may aid in the growth of knowledge relevant to the unambiguous recognition of scavenging versus hunting from archaeological bone assemblages.

CHAPTER **2**

Klasies River Mouth: A Provocative Case

Introduction

The South African region is one of the few areas in the world where there are well-studied faunal assemblages from a wide variety of archaeological sites. This remarkable and admirable situation is largely the consequence of the dedication of a single researcher, Richard Klein, of the Department of Anthropology, University of Chicago. For over a decade Klein has been developing a faunal "library" that is essentially unparalleled elsewhere. A significant result of these efforts has been Klein's recognition of some interesting forms of patterning, for which he has, quite rightly, offered interpretations.

Because of Klein's work, there has gradually emerged a procedure for describing and studying fauna, as well as an interpretative model distinctly associated with Klein's view of the past. Looking through Klein's publications from 1972 to the present makes an absorbing exercise; in them the development of his analytical tactics is well illustrated. In addition to a development of analytical strategies, there is an accumulating recognition of patterning in different properties of faunal assemblages and attendant building up of interpretative arguments. I am particularly interested in a pattern that Klein (1974) recognized and first described from his analysis of the Klasies Middle Stone Age (MSA) materials:

> some possible limitations on their hunting capabilities may be implied by the fact that, in contrast to later peoples in the area, they concentrated their attention on the

> most docile of the available large bovids (eland) and largely ignored the (? too dangerous) suids, one or both species of which were probably abundant in the vicinity. It is further interesting that the eland remains belong overwhelmingly to adults, while the other large bovids—the buffaloes—are represented to a large extent by young to very young animals. The cunning and ferocity of the buffaloes (inferred for the extinct giant form) would have made the adults exceedingly dangerous prey. (Klein 1974:270)

This general picture of the patterning observed for the Klasies River Mouth fauna is reiterated (Klein 1975b) with the added suggestion that hunting by man might be a contributing factor to the extinction of some African species.

During the initial evcavation and preliminary reporting phase of the work at Klasies River, several points were made that were of extreme importance. The excavators saw no evidence for major interruptions in the use of the site spanning very long periods of time, leading them to suggest that occupation had been essentially a "permanent settlement." "For hunting–gathering communities to live close to each other and apparently for generations to continue doing so for so great a period of time suggests unique conditions" (Wymer and Singer 1972:209).

Aside from suggesting a kind of MSA sedentism in a coastal "Garden of Eden" (see Binford 1983b for a criticism of this view), this site yielded rather unequivocal evidence for the regular exploitation of aquatic resources (see Figure 2.1 for location). One of the common generalizations found in the literature prior to Klein's work at Klasies River was that the systematic exploitation of aquatic resources was a phenomenon that came

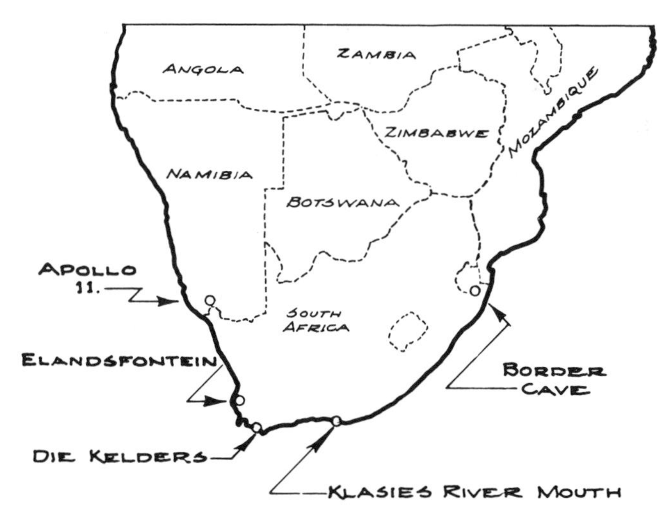

Figure 2.1 Middle Stone Age sites in southern Africa.

late in human history, and was largely characteristic only of the Late Pleistocene (or *Upper Paleolithic* in European terms). Data presented by Charles McBurney (1967) hinted at man's earlier use of aquatic resources, but it was the evidence from Klasies River Mouth that forced the recognition that early man was using aquatic resources for a long period of time prior to the Late Pleistocene.

One of the interesting debates surrounding the Klasies River Mouth sites concerns the age of the deposits. Early estimates of age were largely dependent upon the accepted chronological scheme at the time, coupled with ^{14}C dates. Reasoning from this perspective, the excavators suggested approximately 50,000 years for the span of hominid use of the locations. Later studies by Butzer (1978), using more specialized techniques and linking the geomorphology and sedimentary data to an understanding of sea-level changes, suggested that the MSA occupations were between 70,000 and 125,000 B.P. Work by N. J. Shackleton (1982), using oxygen isotope analysis, estimates the span of the MSA assemblages as ranging between 30,000 and 130,000 B.P. This span seems to be more in line with other chronologies, and would place the shift between MSA and Late Stone Age (LSA) in the South African region at a point in time comparable to analogous shifts to Upper Paleolithic types of material remains in other parts of the world.

The uncertainties of the chronology have, however, contributed to a further problem. Fragmentary hominid remains were recovered from the Klasies sites (Singer and Smith 1969; Singer and Wymer 1982). At least one mandible is considered by most to represent fully modern man, yet the current dating of the deposits in which it was found would require us to believe that humans of our type were present in southern Africa some 80,000 years earlier than in other parts of the world (see Beaumont 1980).

Clearly, the evidence of coastal resource use, the dating, the fossil contents, and the base-line position that Klasies currently serves for comparative purposes (see J. D. Clark 1980) make this a most intriguing area, quite independently of its provocative faunal patterning. Given a still-remaining potential for learning as is suggested by the fauna, the Klasies sites take on an even greater interest.

Environments Past and Present at Klasies River

I suggest shortly that the deposits yielding archaeological remains at Klasies River Mouth have been seen primarily as geologically formed units, within the framework of which hominids periodically appeared and carried on certain activities on the surfaces of the geologically defined deposits. This means that the overall framework, in terms of which the hominid actions may be set into the past, is an environmental one based on inferences from the natural-formation contexts that enclosed the archaeological materials. These contexts were, of course, the dynamics of both the gross and the microenvironments at the site. A knowledge of the potential dynamic conditions that might impact the particular static deposits seen at Klasies River Mouth furnish the base for reconstructing the past environments, and hence provides the important clues to the ecological setting of the hominids' behavior. In addition, some historical knowledge or belief about changes in past environments has served, and continues to serve, as a basic and important guide for dating the deposits.

Present Environmental Setting

The modern vegetation in the region of the caves at Klasies River Mouth is generally referred to as *fynbos* (see Day *et al.* 1979). This term designates the rather unique type of evergreen Mediterranean vegetation found along the coast and into the coastal mountains of southwestern southern Africa. The *fynbos* biome is comparable to the four other areas of the world where the climate is Mediterranean—characterized by cool, moist winters and hot, dry summers. In the Mediterranean region itself, the vegetation is called *macchia.* Along the California coast, particularly between San Diego and Monterey, the analogous biome is *chaparral.* On the coast of Chile the vegetation is named *matorral;* and in southern Australia the word *heath* refers to a similar biome. This, then, is not the environmental zone that one typically imagines when thinking of Africa. Perhaps it is worthwhile to place the southern tip of Africa in a comparative framework, so that its setting may be more realistically appreciated.

The area at Klasies River Mouth (sometimes called the Tzitzikamma coast) is on the same latitude (about 35°S) in relation to the equator as Myrtle Beach, South Carolina, on the east coast of the United States, and the Santa Barbara Channel of California on the west coast. On the other side of the Pacific, Osaka, Japan, and Sydney, Australia are in similar positions relative to the equator; so is Montevideo in South America. Within the Mediterranean area, the north coast of Lebanon and the North African coast around Casablanca are comparable in climate and distance from the equator.

The Klasies River Mouth setting is not only analogous to California in terms of latitude, but is also very similar in terms of topographic conditions.

Figure 2.2 The southern African Coast, looking east near Die Kelders, South Africa. Note the rocky outcrops and the low *fynbos* vegetation.

First, the coast is rocky (Figure 2.2) and drops off sharply to a deep immediately offshore. This means that fluctuations in the height of sea level could be substantial, but the horizontal displacement of the actual coast from its present position would be at a minimum. At least at Klasies, this indicates that the sea was always relatively close to the cave as the crow flies, but during low-sea-level eras some climbing down would have been required to get there. As with Santa Barbara, there is a coastal mountain range paralleling the coast. At Klasies the mountains crest at an elevation of around 600 m only about 12 km inland from the present coast. These mountains are the eastern tip of the Cape Folded Mountains. The coastal side of the mountains at Klasies River Mouth is really a series of three narrow terraces (Figure 2.3). The first rises abruptly out of the sea to an elevation of 200 m. This elevational notch is made up of coastal cliffs that rise from 60 to about 90 m above current sea level. Variously filling in the convolutions of the "table mountain quartzite," which forms the bedrock of the coastal cliffs and is also the home of the caves that we will be discussing, are vast coastal-dune formations. These generally make up the surface of this first topographic notch as we move back from the coast. Still farther, there is a slightly higher notch with elevations between 140 and 160 m above sea level. Finally, at the base of the mountain range, the last notch is at about 275 m. The base of the mountains is only about 12 km inland from the coast.

Two rivers drain the area between the caves at Klasies River Mouth and the base of the mountains. The Tzitzikamma River picks up most of the runoff from the mountain proper. It has its headwaters to the west of Klasies River Mouth, and drains southeastward along the base of the mountains, entering the sea some 5 km east of the Klasies River Mouth. The Klasies River essentially runs along the junction of the first and second

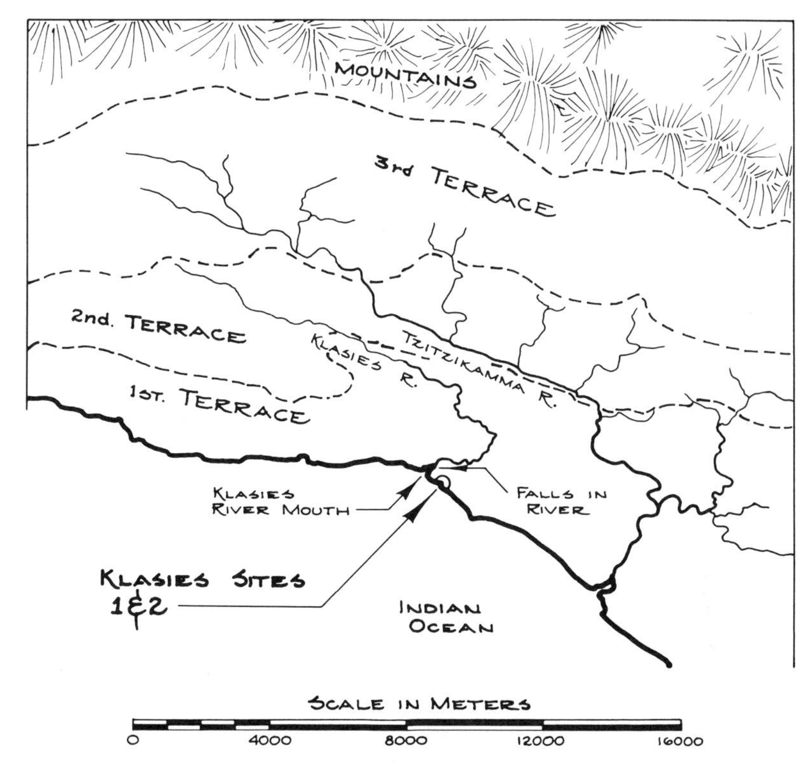

Figure 2.3 The Tzitzikamma coast, showing location of caves and the gross topographic setting at Klasies River Mouth.

notches on the coastal side of the mountains. Its course parallels that of the Tzitzikamma drainage (see Figure 2.3), only at a lower elevation, and turns southwest as it cuts through the first coastal notch, cascading over falls into the presently drowned river gorge, locally called Klasies River Mouth.

Today, Klasies River Mouth is on the eastern margin of the Cape climatic province, which is in its more westerly expression characterized by predominantly winter rainfall. The Klasies River Mouth annually receives about 750 mm of rainfall with a late autumn–early winter maximum, although rain may occur throughout the year. The temperature variation is not great, with a mean temperature of the coldest month (July) of 14.2°C (57.3°F), while the warmest months (January and February) average 20°C (67.7°F). As can be seen, this is a very uniform climate in that it does not suffer great temperature extremes seasonally. This climatic evenness is conditioned by the ocean; the interior areas north of the Cape Folded Mountains are generally excluded from the tempered conditions on the coast.

Vegetation and Ecology

In spite of the fact that rainfall may occur year round on the Tzitzikamma coast, the vegetation is Mediterranean in character. The most striking characteristic of the *fynbos* is that it receives most of its moisture coincident with the period of least solar radiation, and in turn the greatest solar radiation occurs when rainfall is least. The plants therefore are adapted to reduce transpiration during dry, warm seasons, typically having hard-surfaced leaves. Plants maximize the reduced solar radiation during the wet season, being evergreen and yielding a low rate of production. The result is that plants of this biome exhibit a very low turnover rate; that is, the proportion of new cells added is low relative to the amount of cells being maintained from one year to the next. Vegetation biomes such as the *fynbos* have some metabolic properties in common with high-biomass forests, where the total amount of production is relatively low but most of it is maintained as increased biomass stems, twigs, and perennial plant tissues. Fynbos is a high-biomass biome relative to its rate of production for new cells annually. In general, such biomes are poor places for animals to make a living. Animals generally eat new growth and/or reproductive organs of plants—in short, the *production* of a biome. The predominance of plant species for which most new growth is transformed into permanently maintained plant structure (such as branches) ensures that food is sparsely distributed for plant-feeding animals. Environments with high biomass relative to production just do not support many animals; they support more of their own kind instead.

Ethnohistorical information (see Parkington 1972) regarding the uses of the *fynbos* by recent peoples show that plant foods that commonly form a staple of hunter–gatherer existence, particularly in subtropical settings, are relatively rare. In fact, there are only a very few species of known consumable plants. Some of the plants yield seeds that are useful for their oils but provide little else. However, a variety of fuits and berries are produced during the summer months (January and February).

By far the most important food plants are those having edible rootstocks. Two types are particularly useful: those having corms like a garden iris, and those having bulbs like an onion or shallot. In fact, some of the most nutritious of the edible corms are all members of the Iridaceae, or various wild relatives of our garden flower, the iris. Most of these plants reach maximum size during early summer, while the rootstocks are shriveled up during the winter. In general human foods are most available during the summer months, although these are rather sparse except for the corms and bulbous plants, which could supply a summer staple. It is difficult to imagine how hominids could have lived year-round in the coastal *fynbos* without a systematic use of the resident animals and coastal resources.

Grasslands have very different ecological properties. They have quick-turnover ratios, in that very little of the annual production is in fact maintained as standing biomass. These are generally environments where limited rainfall coincides with peaks of solar radiation. Quick-turnover grassland is commonly found on well-drained soil when rainfall is deficient, or on poorly drained soil enjoying moderate rainfall. Although production in the plant community is high, evapotranspiration is also high, so that the ground moisture needed to sustain substantial biomass does not last into the cool season. One reproductive strategy of plants in such environments is to grow quickly, produce many seeds, and then die down to the roots; in this way, very little moisture is needed to maintain the limited biomass in the root system. Another strategy is to die completely but to reappear as a new generation of plants the next wet season, largely germinating from seeds.

One might think of grasslands, then, as characterized by plants that behave like annuals (many are), and high-biomass environments as biomes that behave as if made almost exclusively of perennial plants. Trees and shrubs send their roots deeply to maintain a constant supply of moisture to the aboveground plant structure and to support the greater mass of their structure. This means that the moisture should generally increase with greater depth, needed to ensure sufficient root support for the aboveground plant. This occurs when soil is well drained, particularly if it is sandy. When rainfall occurs, it quickly penetrates and is absorbed into the earth. Evaporation is increasingly prevented as the water percolates to greater and greater depth, eventually meeting the underground water table—which means that moisture-seeking roots are increasingly rewarded as they grow downward.

In contrast, where moderate rainfall occurs during the warm season, there is a race for water to penetrate the substrate before evaporating. Fast-growing plants like grass generally capture moisture near the surface. There they quickly bloom, to die back as the moisture at the surface is lost both to evaporation and transpiration of their own making. Moisture therefore rarely penetrates to the zone between the superfically wet areas and the underground water table. This means that there is a generally dry zone between the surface and the underground water table. This dry zone discourages the roots of trees and shrubs, which find less and less moisture as they penetrate the deeper levels. The result is that grass dominates.

Increases in summer rainfall, so that it far exceeds the potential evapotranspiration as regulated by solar radiation, ensure more and more deep saturation, and eventually an area may be transformed into savanna or, with even more moisture, woodland. This same expansion of trees and shrubs with increased rainfall may be enhanced by increases in winter rainfall, since evapotranspiration is at a minimum and deep penetration of moisture therefore more likely.

The *fynbos* of the southern Cape is a relatively high-biomass biome, maintained by well-drained soils and winter rains. The combination of conditions ensures that moisture penetrates well below the surface, so that the deep-seated roots of the shrublike plants are not turned back by a dry, forbidding subsurface. The presence of summer rain and the further reduction of evapotranspiration conditioned by the climatic evenness along the coast encourages high-biomass vegetation to predominate, as opposed to quick-turnover grasslands. Increases of a winter rainfall regime tend to favor expansion of temperate deciduous forests—in short, biomes of greater biomass. Reductions in rainfall tend to reduce the overall biomass, but not necessarily to favor grasslands unless there is a correlated shift to a summer rainfall pattern.

On the north side of the Cape Folded Mountains, there is a very different vegetation province. Today, the term most commonly used for the biome immediately north of the mountains is *karoo*. There, the vegetation is sparse, with extensive bare areas. Seasonally, depending upon the summer rainfall pattern, there may be grass patches. Trees, restricted generally to

Figure 2.4 "Bushman hunting a herd of heterogeneous game," by T. Baines. The watercolor drawing shows black wildebeest, zebra, ostrich, and what appear to be bastard hartebeest. This is the association of animals found in the *grassvelt* to the north and east of the Tzitzikamma coast.

watercourses, are predominantly acacias. Today the *karoo,* on the interior side of the Cape Folded Mountains, becomes relatively lush to the east, where it tends to merge with the better-watered *grassveld* of the eastern region. The ideal *karoo* is characterized by summer rains and dry winters, and it supports a fauna typified by grazing animals, whose forms are well adapted to dry, barren plains with seasonal blooms of grass. The large-animal fauna of the interior *karoo* consists of springbok, gemsbok, wildebeest, and zebra. As rainfall increases eastward—and in the more lush grassland of the southern Transvaal and what is today Lesotho and the Orange Free State—vast herds of migratory animals were encountered, moving seasonally with the grass. The herds are dominated by springbok, black wildebeest, zebra, and blesbok (Figure 2.4) or what Klein (1976) generally calls the "bastard hartebeest". This is the world of classic African grassland with its vast herds of migrating animals (for a good early description, see Inskeep 1978:12–13). It is quite literally cut off as one approaches the Cape Folded Mountain range, which separates the summer rainfall zone from the coastal winter vegetative zone of the *fynbos* biome. Today the *karoo* is a rather forbidding area supporting very little animal life. Nevertheless, it does represent the low-rainfall margins of the summer-dominated precipitation zone, which in its better-watered areas supports the vast grasslands of subtropical Africa. These, in turn, support the vast herds of grazing animals that we tend to associate with Africa.

Past Environments

In terms of understanding past environments, there have been several models used by archaeologists for interpreting environments of the past. Perhaps the first widely held idea, promoted by Louis Leakey in the 1930s, was that there were *pluvial* periods that alternated with drier *interpluvials.* These were thought to correspond to the glacial and interglacial episodes of Europe and the New World. This view was generally abandoned when it was realized that the climatic history of the Pleistocene was much more complicated than previously imagined.

Beginning in the late 1960s, van Zinderen Bakker (1967) began to popularize a model of contracting and expanding climatic zones. This idea was adopted and elaborated upon by van Zinderen Bakker (1976), Tankard (1976), and Tankard and Rogers (1978). The model was similar to a balloon blowing up and deflating. During glacial periods (Figure 2.5), when sea levels were lower, the modern climatic zones were thought to contract symmetrically toward the equator. During periods of high sea level and interglacial–interstadial conditions, the reverse would be expected, so that

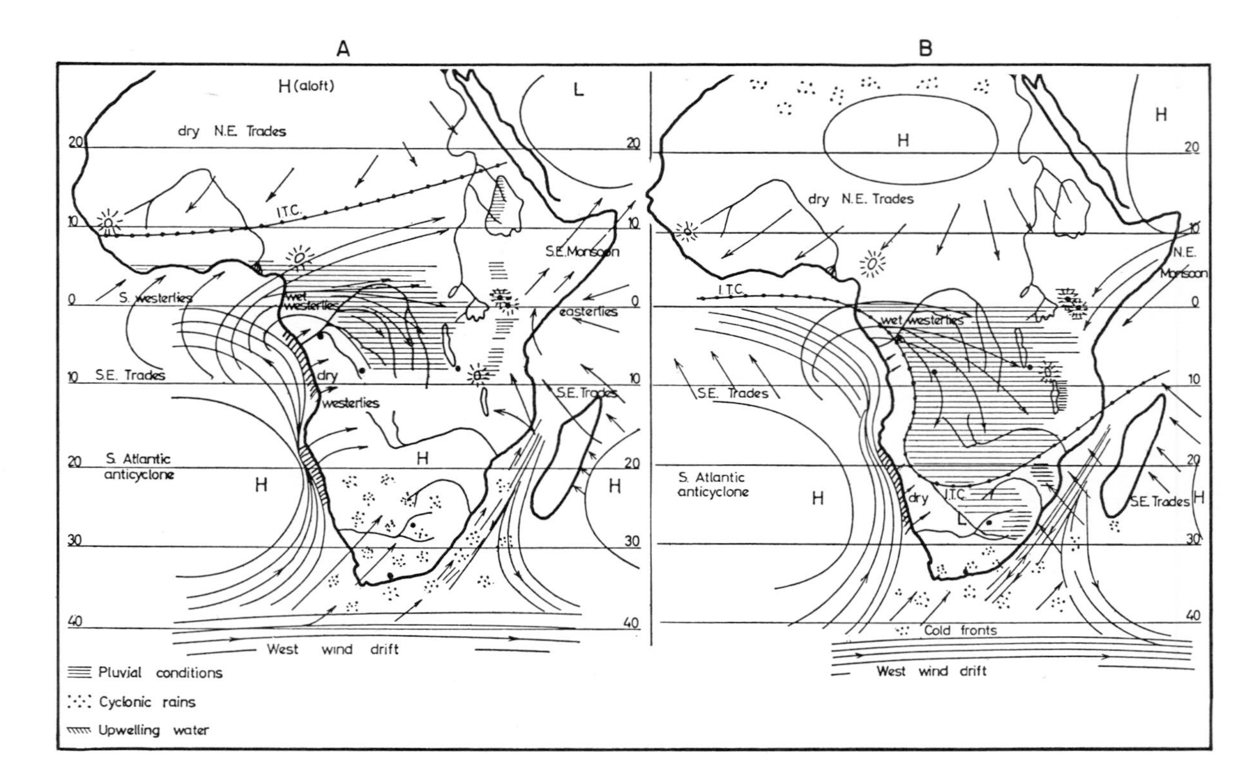

Figure 2.5 Climatic conditions in (A) winter and (B) summer, as reconstructed for a typical glacial episode according to the zonal model of van Zinderen Bakker (1967).

the modern climatic zones would expand outward from the equator (Figure 2.6). The picture that emerged from this model was for there to be a movement north (toward the equator) of the winter rainfall pattern during glacial maxima, with a corresponding shift even closer to the equator for the zone of summer rainfall. Given the description of environments presented here, we would therefore expect more winter rainfall during glacial episodes on the interior plateau beyond the Cape Folded Mountains, where today there is marginal *karoo* along the edges of the summer rainfall zone (see Figure 2.5). Correspondingly, we would expect expansion northward of summer rain-regulated grassland at the expense of forested areas nearer the equator.

This model, if it is correct, should permit us to anticipate some of the environmental conditions along the southern African coast: (1) during periods of glacial maxima coincident with lowered sea level, the *fynbos* should expand beyond the Cape Folded Mountains as winter-dominated rainfall patterns shift northward. Correspondingly, grazing animals should be least likely to occur in coastal sites, which should be dominated by the browsers and mixed feeders. (2) During interglacial–interstadial periods coincident with raised sea levels, the *fynbos* should be pushed back south of the Cape Folded Mountains. The *karoo* of today should become more lush, with grazing animals most likely on the south coast and the browsers and mixed feeders becoming least common.

Interestingly enough, Richard Klein (1972) has published a most intriguing body of data from Nelson Bay Cave on the southern coast just to the west of Klasies River Mouth. Importantly, this site is located on the very edge of one of the more interesting temperate-forest biomes known on the coast. In recent times, at least, the relatively high biomass in the area was made possible by the well-drained soils and the winter rains. Most of the deposits at Nelson Bay Cave are within the reliable range of ^{14}C techniques.

Historically, the fauna recorded there were the Cape buffalo (Figure 2.7) bushpig (Figure 2.8) bushbuck, and grysbok, all browsers or strongly mixed feeders, as is the Cape buffalo. These are the animals expected in a high-biomass, metabolically slow biome. Yet the deposits dating between 18,500 and 12,000 years B.P. corresponding to the end of the glacial maxima at the close of the Pleistocene proper—demonstrate that the fauna was in fact largely made up of grazing antelopes and equids: zebra, wildebeest, springbok, and bontebok (a southern analogue to the topi of East Africa).

Here then, we see situations directly opposite to the conditions anticipated by the zonal model that predicts that during glacial maxima there would have been an expansion of the high biomass from the coast toward the interior; and that during interglacial–interstadials, the interior summer rainfall pattern would have expanded southward. This would mean more rain in areas that today are *karoo*, and perhaps even a summer rain pattern

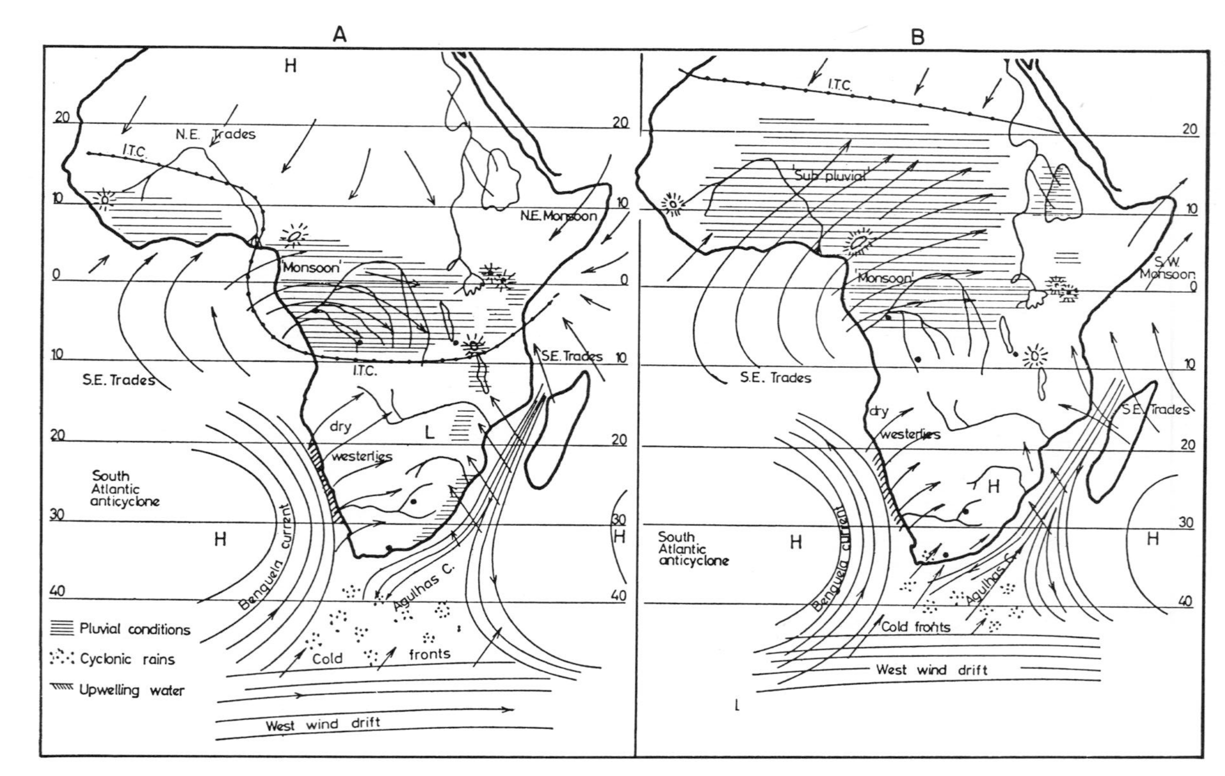

Figure 2.6 Climatic conditions in (A) winter and (B) summer, as reconstructed for a typical interglacial episode according to the zonal model of Van Zinderen Bakker (1967).

Figure 2.7 Cape buffalo in typical bush cover. (Courtesy of J. Ebert.)

in areas that today are transitional, such as the Klasies region. Farther west, the lower rainfall of the *karoo*-like zone would be expected along the coast. We would expect browsing–mixed-feeder faunal associations during glacial maxima and grazing animals during interglacials. But the facts are just the reverse: we have grazing animals during the glacial maxima, and browsers and mixed-feeders in historic times associated with a high-biomass temperate forest.

Figure 2.8 Bushpig in bush cover. (Courtesy of J. Ebert.)

Based on sedimentological argument, it has been suggested that during cold intervals rainfall was reduced on the coast along the western Cape, with the suggestion that this favored grassland! On the other hand, along the eastern Cape there is evidence that rainfall increased during cold periods, permitting the expansion of deciduous forests into areas where today *fynbos* is found (Schalke 1973). This sustains the picture given by Klein's data on fauna.

The "reverse" zonal model is further supported by data from Elands Bay Cave (Butzer 1979; Parkington 1980, 1981; Miller 1981), where there is a substantial grazing fauna associated with cold conditions, low sea levels, and drier settings—in short, a glacial condition at the close of the Pleistocene. The paleoenvironmental data as summarized from fauna suggest that the zonal model is at least partially correct, but *backward* as far as southern Africa is concerned. During periods of glacial maxima (low sea level and colder climate), the interior summer rainfall regime apparently expanded southward, favoring grass along the east coast and reduced rainfall, perhaps of a summer pattern much like parts of the *karoo* today. Alternately, during interglacial–interstadial conditions (corresponding to warmer sea temperature, higher sea levels, and warmer climate), the summer rainfall pattern contracted northward, yielding a pattern similar to that of today, and at times perhaps even moved farther northward of the winter rainfall pattern than is known today.

The disconcerting implication of this situation is that interpretations of glacial versus interglacial conditions surrounding the accumulation of deposits in which archaeological remains are found can be made using either the original zonal expectations of the van Zinderen Bakker model (Figures 2.5 and 2.6) or its modified version, which expects climates to be the *exact opposite,* based on faunal sequences and their implied climatic conditions! Much of the presently accepted comparative chronology for MSA sites in South Africa reflect these ambiguities.

Stone Assemblages of the Klasies River Mouth Sites

The Klasies River Mouth has been well-reported previously (Butzer 1978; Klein 1974, 1975b, 1976, 1979, 1982; Singer and Wymer 1982; Voigt 1973a,b; Wymer and Singer 1972). The sites are a complex of rockshelters and small caves arranged along a V-shaped draw that rises approximately 23 m from its low point at the mouth of the main cave (Cave 1) to the top of the remnant deposits at the head of the draw in Shelter 1A (Figure

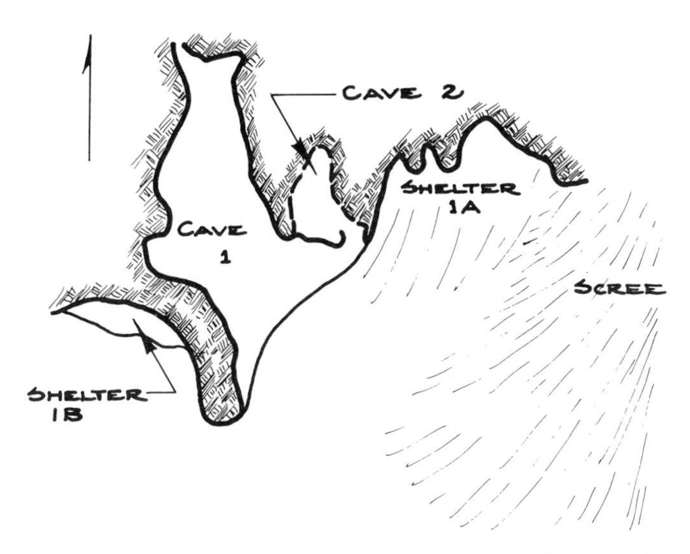

Figure 2.9 The relationship among the various sites at the main site at Klasies River Mouth.

2.9). Before one enters the draw from the west there is a small shelter directly above the present storm beach on the modern coast at roughly the same elevation as Cave 1; this is Shelter 1B. Finally, a "second story" cave is located in the face of the rock within the draw, essentially above Cave 1 and between it and Shelter 1A at the head of the draw. Today, there is a sheer rock face below the mouth of Cave 2, making entry difficult. However, it is believed that when it was occupied a talus deposit in front of the cave mouth ran along the rock face to form a continuous deposit with that of Shelter 1A. For ease of discussion we may speak of Shelter 1B and Cave 1 as the lower caves, Cave 2 and Shelter 1A as the upper caves (Figure 2.10). In both the upper sites there are dark humus lenses and ash deposits, typical of what can be expected in bedding areas and their associated bedside hearths, as well as more communal cooking features (see, for instance, the deposits described by Walton 1951). Both the upper sites have deposits more commonly found in small habitation sites known from relatively recent times. These ash–humus lenses typically dip down along a rock wall in a fashion similar to bedding areas (such as the excavated bedding areas at the site of De Hangen [Parkington and Poggenpoel 1971]).

These ash and humus deposits yielded an assemblage known in the South African literature as Howieson's Poort. This is particularly interesting because it has been generally considered the earliest Upper Paleolithic type of industry in southern Africa. It is characterized by a general reduction in

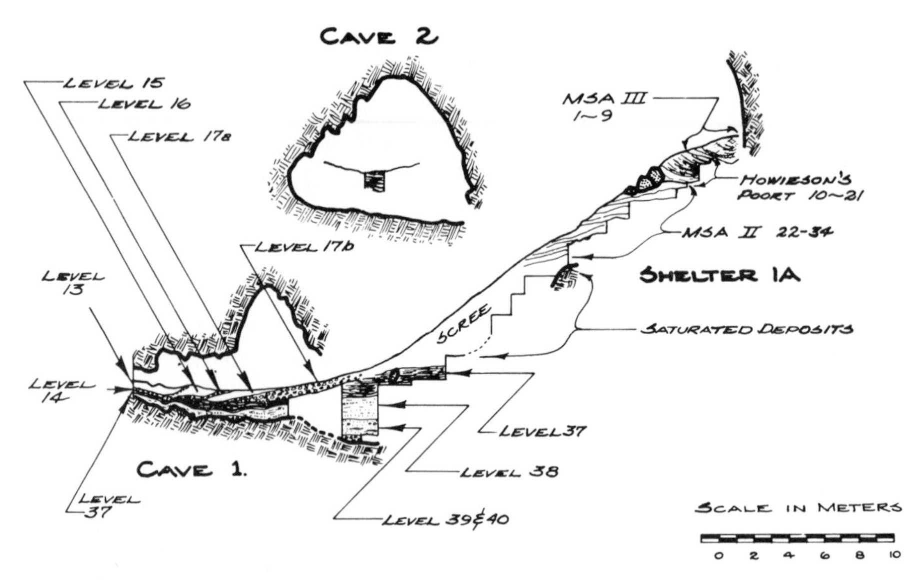

Figure 2.10 Vertical section showing the relationship among the various groups of levels at the sites of Cave 1, Shelter 1A, and Cave 2 at Klasies River Mouth. (Redrawn from Singer and Wymer 1982:Figure 3.1.)

the size of tools, an increased use of raw material exotic to the site area, and the systematic presence of crescents (*backed pieces*, in European terms) and related forms of worked-blade section. The latter is seemingly correlated with the appearance of a punch-blade technique of blade production. As in other Upper Palaeolithic assemblages, there is a marked increase in evidence of personal ornaments in the form of ostrich eggshell beads as well as a correlated increase in pigments, particularly red ocher (see White [1982] for a discussion of the Upper Paleolithic transition).

Equally fascinating was the sequence of levels both underneath and above the Howieson's Poort materials in Shelter 1A. Below were a series of levels yielding an industry typically called Middle Stone Age. This assemblage yields very few shaped tools but large numbers of well-controlled flake-blades, generally struck by direct percussion from various forms of blade cores. The result is usually called a *pointed flake*, with sharp converging or parallel sides. Technically it is a blade because it is twice as long as it is wide. Not uncommon are worked *points*—well-shaped flake-blades that have received some degree of lateral or basal retouch, rendering them sometimes very pleasing. These are very similar to the *leilira* that have been observed in manufacture by the central desert Australians (see Binford and O'Connell, in press; Spencer and Gillen 1972), for whom they have served as knives, spearpoints, and were edged for picks. Among the other shaped tools are scrapers, which are simply flake-blades with retouch somewhere along the margin, and denticulates, which are mostly microdenticulates in a Bordean system of classification. The latter are simply scrapers in which the retouch is not continuous but is characterized by a series of separate chip removals. Finally there are flake-blades that have been utilized but are not judged to have been retouched. They account for around 40% of the shaped tools, with scrapers and denticulates accounting for about 50%. Denticulates are generally more common by a ratio of 1.5 to 1. The remaining 10% is made up of worked points, burins, and borers. In other MSA sites the Levallois technique has been reported, and handaxes are occasionally reported from MSA contexts. Varying degrees of bifacial retouch and modification have been employed for producing sometimes surprisingly symmetrical tool forms. Both the patterns of regional and chronological variability are at present poorly known (see Sampson 1974 and Volman 1981).

If one stands way back and looks at the character of tool assemblages in terms of design properties as they are known to vary with tool functions among ethnographically documented stone tool technologies, the MSA appears to be only minimally differentiated functionally. For example, we know that almost all the tools and containers used by the Australian aborigines were manufactured from wood. We also know that, at least in the central Australian desert, aborigines did not dress hides and they generally did not skin animals. Cutting tasks were minimal and generally were performed with wat we would typologically call utilized flakes. In a similar, very general sense the stone technology is dominated by steep-edged tools commonly used as adzes and woodworking tools. Coupled with this were various rather makeshift core tools, which are large and chunky and were used in obtaining wood (see Hayden 1979). In a very real sense the technology is a scraper–adze-dominated industry, with varying amounts of heavy-duty expedient tools; cutting tasks were generally carried out with unspecialized flakes.

By way of contrast, in the Great Basin of North America all the containers were manufactured of basketry, and clothing was made from either plant fiber or cut-and-woven small-animal skins or bird feathers. Skinworking was at a minimum. Some woodworking was needed to produce throwing sticks, digging sticks, and bows, but wood was a rare part of the technology, compared to the Australian tool kit. Most of the stone tools were used as portable weapons or for cutting tasks. The result is a tool assemblage that is dominated by projectile points and various forms of utilized flakes, which apparently were the most common cutting tools.

Recent studies by H. J. and J. Deacon (1980) have shown a strong correlation between small (hafted) convex scrapers and the use of leather clothing in Africa. The Deacons' study confirms a long-held, intuitive understanding of the role of such tools in technologies—for instance, on the American Great Plains and in the Arctic. In both places, dressed skins were very important clothing items, and there also we find the small convex scrapers as a major part of tool assemblage.

I do not wish to imply that we understand in functional terms technologies in general. There are, however, some very broad patterns against which to compare technologies. The MSA does not appear to have extensively used stone tools for woodworking or for skin-working. Almost all the tools are most consistent with a variety of cutting and scraping tasks. If one views collections of stone artifacts with substantial retouch and secondary and tertiary modification as giving clues about the *use life* of tools (that is, how long tools were actually either planned for use or were actually employed as tools), then the MSA assemblages must rank as very expedient, because tools had very short use-lives. One obtains the picture of a very regularly produced and expeditiously used toolkit, most generally employed in cutting tasks of moderate-to-light duty. Expedient production—that is, tools made for immediate use and then discarded after each task—is indicated not only by a lack of retouch but also by the rather surprising quantities of stone tools and debris relative to the numbers of bones or shells associated in archaeological levels. In my experience, this characterizes pre–Upper Paleolithic sites in general, even when preservation of faunal remains is quite good.

In contrast to the remarkably dull and uninteresting character of the MSA lithics, the stratigraphic distribution of the typologically MSA materials relative to the more interesting and complex Howieson's Poort variant is truly fascinating. We have at Klasies River Mouth a classic case of "alternation of industries (see Figure 2.10); there is no gradual shift in the form of assemblage content among stratigraphically adjacent levels. In Shelter 1A, Levels 36–22 yielded a MSA assemblage (MSA II). Directly *above* these were Levels 21–10, which yielded a Howieson's Poort assemblage. Then resting stratigraphically on top of the Howieson's Poort levels were Levels 9–1, which yielded an MSA assemblage again (MSA III). Stated another way, the stone tool assemblage from Levels 1–9 were most like these from 22–36, and least like those from Levels 10–21. This is a classic case of alternating industries, as noted by F. Bordes (1961) for many Middle Paleolithic sites in western Europe, and as is also known from Middle Paleolithic sites in the Near East (see Binford 1982b; Jelinek 1982).

Because the MSA assemblages are rather unspecialized and generally lack temporal specificity, it is difficult to relate chronologically the deposits from the lower caves and shelters to those from the upper sites. In the case of Cave 1, there was a remnant deposit, which represented a cone formation that had originally accumulated at the mouth of the cave and trailed off into the interior of Cave 1, both toward the rear and to the west (see Figure 2.10). On the west, the cave's floor and roof converged, leaving little available living space. In none of the remnant levels was there any indication of bedding areas as characterized the deposits from the upper shelter, 1A. The excavators noted no localized fires or features that could be recognized as modifications of the surface produced by hominids.

The layers of Cave can best be viewed as the result of geomorphological formation processes, with the hominids fitting their activities onto the surfaces as they were modified by such natural processes as erosion and physical accumulation. After an initial period of rather intense use, resulting in humus-laden occupational soils and ash lenses near the mouth of the cave where the deposition was relatively level (Levels 38 and 37), the formation of a major cone at the mouth of the cave was started. This is believed to be at least partially the result of an encroachment of the massive fan deposit of scree that forms the present east side of the draw (Figure 2.9), coupled with tumbled deposits coming down from a secondary fan of occupational debris centered in Shelter 1A above. Items occurred in this level buried at all angles, suggesting that most of this rubble was a secondary deposit. This possibly represents rubble deposition that formed the base of the MSA II deposition in Shelter 1A. This speculation is simply based on a projection of the bedding angle as illustrated in Figure 3.1 (Singer and Wymer 1982:10) of the original site report.

Level 16, which rests on top of the angled rubble deposits of Level 17, marks a period of sand encroachment into the cave, believed to represent blown-in beach sand. Shell was noted as particularly common in this level, and there was at least one localization of burned stone and bones near the cave mouth, presumably marking an active, *in situ* hearth. Above this sandy level was another layer, Level 15, which had numerous ash lenses and evidence of *in situ* accumulations of both tools and food debris. On top of this cultural level rests another rubble level, Level 14, which is very important because it is quite rich in archaeological remains. Unfortunately, it is difficult to understand in depositional terms. It is likely that this is mostly redeposited from tumble or scree accumulating from the expanding fan of scree centered just in front of Shelter 1A. It also appears that this deposit was further modified in Cave 1 by a high-sea-level stage, during which the cave apparently was periodically washed by very high storm-tides. It is my guess that Level 14 is derived partly from conditions surrounding the fall of large blocks at Shelter 1A, which rest on Level 22 (Shelter 1A), and partly from the conditions represented by the high rubble concentrations of Level 22 itself.

Although other deposits occur in Cave 1, the final level of great interest to us is Level 13, which has been interpreted as mainly accumulated from wind-blown sand. Laced through the sand deposit were small lenses of silt and clay. No hearths were observed, and no soil development that could be referred to the accumulation of organic debris from hominid occupation was noted in the deposit. This level has been interpreted as "a typical regressional eolianite, recording a falling sea level that was initially near the cave but ultimately quite distant. A major glacial–eustatic regression is indicated" (Butzer 1982:39).

Several main observations might now be made in this discussion:

1. Although hominids certainly played some role in the accumulation of these deposits, the manner of excavation, preservation, and documentation renders any specific understanding of the way the hominids used the land surfaces of Cave 1 a matter of speculation.
2. The recognized levels seem to be clearly geological in origin, and the contribution of hominids must be seen as contained within the geologic events. That is, the hominids moved across the surfaces that were being created by noncultural processes, contributing to the context of the deposits but not visibly altering the ongoing geological depositional processes.
3. In no sense can we use the excavated data from Cave 1 to discuss the formation processes of the archaeological record, because the archaeological record is only documented in its formal or compositional sense and not in its site-structural or depositional sense.

Any progress that we may make in furthering our understanding of the past will be made, at least from the excavated materials of Klasies River, through a study of the populational characteristics of the materials included in the deposit considered relevant to the behavior of the hominids habitants, studied in relation to each other and to the history of climatic events providing the context for the archaeology.

Dating the Events Documented in the Klasies Sites

In 1970, Richard Klein published a very important paper that sought to summarize understanding of that time regarding the MSA. This paper was important because it was written just before there was a major revolution in the historical views on the MSA. At the time of Klein's paper, he generalized that "the available evidence suggests that the Middle Stone Age of southern Africa was broadly contemporaneous with the Upper Paleolithic of Europe" (Klein 1970:123). Shortly after his review article, a revolutionary time scale for the archaeological chronology of southern Africa was proposed by Vogel and Beaumont (1972) and Beaumont and Vogel (1972). They suggested that the boundary between the MSA and the LSA was around 37,000 B.P. This is approximately the boundary between the Upper and Middle Paleolithic as known in Europe, thus making the MSA of southern Africa not contemporary with the Upper Paleolithic, as generalized by Klein and others, but with the Mousterian of Western Europe. It was correspondingly suggested by Beaumont and Vogel that the MSA began sometime earlier than 100,000 B.P.

This chronological revolution carried with it interesting implications, because the hominid remains recovered from MSA sites available for study at the time of Klein's 1970 review were acknowledged to be poorly documented. However, most were considered consistent with the view that the makers of the MSA tools were fully modern hominids (see Klein 1970:132).

Only shortly after Beaumont and Vogel advanced the revised chronology for the MSA, Butzer argued a compatible early interpretation from sedimentary and sea-level data as observed at Klasies River Mouth. Between 1969 and 1973, Butzer (1972, 1973) was engaged in studies of the southern African Cape with a focus on the chronostratigraphy of the sites at Nelson Bay and Klasies River Mouth. Although not published until later, by 1973 the view was already widely circulated that the Klasies site dated back to what in the French chronology was thought of as the Riss–Würm interglacial of approximately 125,000 B.P. This inference was made based on Butzer's interpretations of ancient sea-levels at Klasies River Mouth. These were matched to the oxygen-isotope curves derived from deep-sea cores considered as particularly sensitive indicators of glacially related ocean temperatures. The resulting chronological scheme has been widely published (Beaumont *et al.* 1978; Butzer 1978, 1982; Klein 1976). The interpretation was rather simple: (1) It was argued that Cave 1 settlement was generally earlier than that of Shelter 1A. It should be pointed out that there is no stratigraphic justification for these assumptious (see Figure 2.10). At no point was the stratigraphic sequence from these sites physically interdigitated. This means that the chronological cross-correlation was an inference from other types of nonstratigraphic data. (2) It was assumed that the caves and shelters had their origins in marine episodes, corresponding to the sea's erosional notching of the coastal bedrock. Butzer (1978) equates the erosional production of the lower caves (Caves 1 and Shelter 1B) with the +7-m sea level considered characteristics of the marine isotopic Stage 5e, as defined by Shackleton and Opdyke (1973). Given this beginning point for the accumulation of deposits in the caves, Butzer assumes that the stratigraphic accumulation in the caves and shelters began shortly after the formation of the lower caves and that the filling was an accretional process leading toward the present. The physical stratigraphy was additionally evaluated in terms of sand sizes, evidence for active sea penetration of the deposits, and variations in water percolating through and on the deposits.

In speaking of Cave 1, Butzer (1978:144–145) states:

> (i) KRM1-40 represents a regressional cave deposit that is only slightly younger than a long-term sea-level of about +7m. (ii) Levels 17 and 16 include lenticles of typical foreshore eolian sand, and artifically introduced beach cobbles are common in 16. This argues for a relatively high, if oscillating sea level, with a sand beach. . . . (iii) Level 14 coincides with a rising sea-level that brought storm-wave action directly into the cave. . . . It is unlikely that the responsible level was more than 4 or 5 meters above that of the present. . . . (iv) Level 13 is a typical regressional eolianite, recording a falling sea level that was initially near the cave but ultimately quite distant. A major glacial-eustatic regression is indicated.

This summarizes the stratigraphy of Cave 1 as it relates to the MSA. It is clear that all the deposits are considered referable to high-sea-level episodes except Level 13, which is identified as having accumulated as the sea was retreating. All the deposits in Cave 1 below Level 13 suggest high sea levels; there are no deposits recognizable as accumulations between periods of lower sea levels. In short, there is no evidence for interruption in accretional accumulation of deposits during low-sea-level conditions.

Nevertheless, Butzer assigns each level beginning with Level 40 to successive phases of high ocean temperature, beginning with Stage 5e (about 125,000 B.P.) Oxygen-isotope curves from the Pacific deep-sea core reported by Shackleton and Opdyke (1973). This is done in spite of the fact that each deep-sea, warm substage alternated with a low or colder substage. As pointed out, there were no depositional facts referable to these colder–lower sea levels that presumably alternated with the warmer substages on the deep-sea cores.

Butzer is fully aware of this, because he points out that, "In effect, all the deposits other than travertines are related to relatively high sea-levels" (Butzer 1978:147). Citing supporting evidence from the form and contents of deposits, Butzer (1978:147) characterizes the Klasies sequence in this way: "Marine shells of the littoral or sublittoral zone are abundant in most KRM 1 horizons, except level 40 (due to partial decalcification), and middle and upper level 13 (here there are land snails but no marine shells.)" These observations, linked with sand-size data, all pointed to evidence that deposits accumulated adjacent to high-water beaches (except Level 13, which, as Butzer has pointed out, probably represents a major lowering of sea level).

Turning to the interesting problem of the relationship of the depositions in Shelter 1A to those in Cave 1, Butzer (1978:147) notes: "In KRM 1A shell generally is poorly preserved, but absolute shell quantities also are relatively low." Butzer then goes on to comment that dolphins are absent, except for one questionable specimen from Shelter 1A. On the other hand, they were relatively common in the Cave 1 deposits. Both of these facts suggest to Butzer less of a coastal marine emphasis in the 1A deposits. This tends to support the view that the Shelter 1A levels were accumulated over a slightly different span of time than the mass of Cave 1 deposits.

It does not appear that Butzer studied the deposits from Shelter 1B, since they are not mentioned in either of his papers (1978, 1982) on the sites. In the final report on the site, however, Singer and Wymer record the following interpretation: "These layers constitute a straightforward succession of occupational deposition commencing on the shingle of the 6–8 m raised beach and thus almost certainly relate to the lowest levels of Cave 1, layers 37–40" (Singer and Wymer 1982:25).

This mechanical inference of contemporaneity with the lower levels of Cave 1 does not consider Butzer's inference of a storm beach represented at Cave 1 in Level 14. There is no reason why the "beach deposit" in Shelter 1B could not have been scoured at the time of wave action or some other high-water event in Level 14 of Cave 1. Content analysis does not seem to have contributed to the chronological inference by Singer and Wymer. However, it should be pointed out that, like the deposits in Shelter 1A, Shelter 1B lacks dolphins, and buffalo of both Cape and giant varieties are not numerous, although Cape seal is very common.

These contents plus other evidence—high frequencies of double-platform cores relative to irregular and single-platform cores, and the very low frequencies of artifacts manufactured from indurated shale (which dominates the assemblages from Cave 1) as opposed to high frequencies of quartz that dominate the Howieson's Poort industries of Shelter 1A—all lead one to be very suspicious of the chronological inferences made for Shelter 1B. Such comparisons, while not conclusive—because we do not know what conditioned the properties being compared—suggest that caution should be exercised when considering Shelter 1B as contemporaneous with the early levels of Cave 1.

If the Butzer chronology is correct, as well as the arguments about environmental dynamics presented earlier, there are certain conclusions that may be reasonably anticipated.

1. If all the levels represent warm (high-water) stages, the environments on the coast should be various manifestations of the present winter-dominated rainfall pattern, with its correlated high-biomass environment (*fynbos*-to-temperate forestlike biomes) supporting primarily browsers and mixed-feeders.

2. If the sequence of high-water stages are correctly equated to the deep-sea isotopic events by Butzer (isotopic Stage 5d–5e = Levels 40–37;

Figure 2.11 Red hartebeest: examples of grazing animals.

isotopic Stage d = Level 16–17; isotopic Stage late c = Level 15; isotopic Substage early 5c = Level 14; isotopic Stage 4 = Level 13), and because deep-sea cores demonstrate a progressive cooling of the seas during the early part of the Upper Pleistocene (Substage 5, in which all the Klasies levels are assigned), We would expect to find evidence of two occurrences: (1) warm, and hence high-water, substages should be progressively lower during the sequence of Late Pleistocene events, and (2) there should be an overall trend in favor of increasing numbers of grazing animals (Figure 2.11) over the span of Late Pleistocene events documented at Klasies River Mouth. In short, we should see a progressive trend of increasing grazing animals with each successive high-sea-level substage until they dominate the remains in Level 13, which is designated a low-water substage, and hence a cold period.

The observed pattern, however, is nothing like this. First, the level of high-water substages is apparently considered to be nearly equal to the high-water stage represented by the interglacial that we are now experiencing. How, then, do we account for the clearly progressive trend in lowering ocean temperatures recorded in the deep-sea cores, particularly if the lower temperatures are thought to correspond to periods of glacial activity? More important, however, the numbers of grazing animals (Figure 2.12) as defined by Klein are dominant in Levels 37–38 and decrease steadily until Level 14, where they start to increase and then increase markedly in Level 13, which is identified as a low-water substage. The increase in grazing animals associated with Level 13 is what we would expect, given what we have seen at places like Nelson Bay Cave and Elands Bay Cave, where

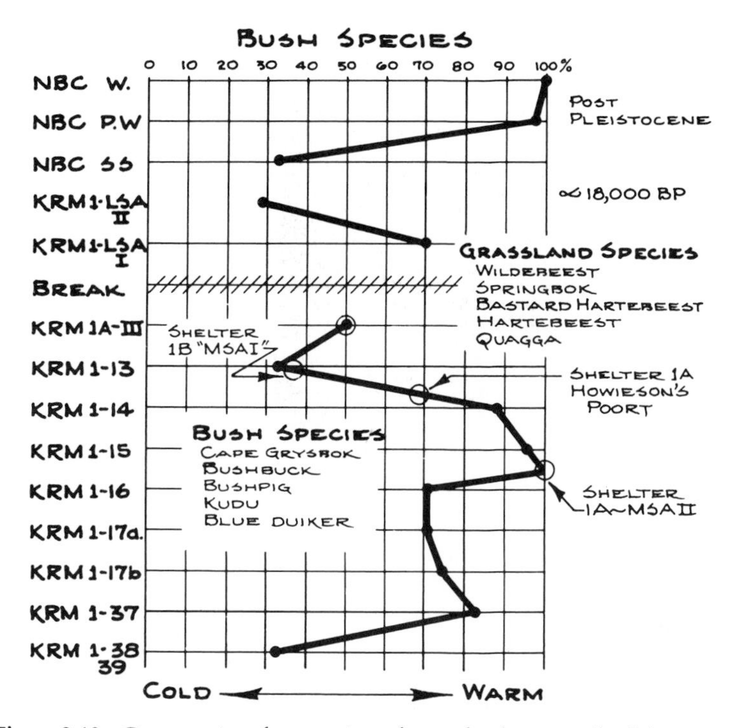

Figure 2.12 Comparative frequencies of grassland versus bush-loving species across the major excavation units at Klasies River Mouth. (Data from Klein 1972, 1976.)

grazing animals are most common in low-water, cold stages. What is not consistent, however, is the apparent decrease in grazing animals across the isotopic high-water substages identified as 5d–5e, running through to 5c in Level 14. It is true that during high-water stages (nonglacial) we should expect increased frequencies of browsing animals coincident with expansions of *fynbos*–temperate forests in the Cape zone, as has been the case in post-Pleistocene times.

Figure 2.12 illustrates the percentage of cover-loving animals (bushpig, grysbok, bushbuck, mountain reedbuck, oribi, vaalrhebok, kudu) versus grass-loving animals (blue antelope, bastard hartebeest, and wildebeest, springbok, hartebeest, and quagga). Based on the fauna, we would expect each successive high-water stage to have been progressively warmer and more dominated by increasing amounts of winter rainfall until Level 14, where the onset of cold conditions would be indicated, reaching a maximum in Level 13, which is acknowledged to be a low-water stage.

It cannot be overemphasized that this overall pattern is at variance with the more general views of climate change for the early Upper Pleistocene:

"The fact that grazer dominated fossil faunas should be so common probably reflects the fact that climatic conditions cooler than the present occupied much more of the Upper Pleistocene than conditions similar to present" (Klein 1980:257). Yet, at Klasies "grazer" species decreased as the glacial environments were, it is thought, becoming more severe.

Clearly there is a problem here. It should be pointed out that in terms of fauna, the contents of Shelter 1B fit into this sequence either between Levels 17 and 16 or, more likely, contemporaneous with the increasingly cold conditions favoring grass-loving species between Levels 14 and 13, making it roughly contemporary with much of the upper Shelter 1A sequence. In other words, the faunal facts support our earlier skepticism regarding the correct chronological placement of the deposits at Shelter 1B.

Given the discussion of the deposits at Klasies relative to the isotopic substages known from deep-sea cores, it was only reasonable to study a number of shells from the Klasies deposits in terms of the $^{16}O/^{18}O$ ratios. Such studies were actually carried out by N. J. Shackleton on two shells from LSA levels in Shelter 1D (not previously discussed here), which yielded ceramics and are believed to date around 100 B.C. (Singer and Wymer 1982); one shell (broken) from Shelter 1A, Level 20; seven shells from Cave 1, Level 15; and one shell from Cave 1, Level 38. One additional shell was studied from Level 12, Shelter 1B; and three others from Cave 5 (not discussed here), Level 6. The latter four shells were all interpreted as MSA, though coming from separate sites that are not stratigraphically juxtapositioned.

Figure 2.13 summarizes the range and the mean for standardized $^{16}O/^{18}O$ ratios between the two most extreme values (summer versus winter temperatures) for each sample studied. In general, the higher the positive

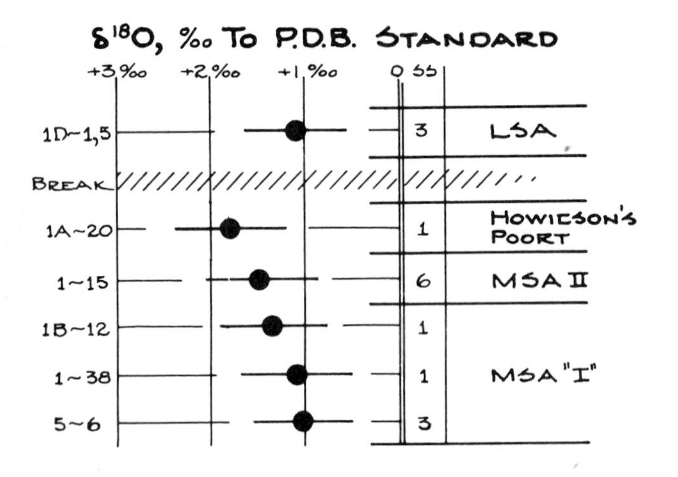

Figure 2.13 Summary graph of $^{16}O/^{18}O$ ratios as known for shells recovered from various levels at Klasies River Mouth. Values plotted are means of the high and low values from each sample and the median value for each. SS = sample size. (From Shackleton 1982.)

deviation from zero, the colder the seas during the period of shell growth. What is very clear is that the LSA shells have a relatively "warm mean value" right at +1. These shells would record the warm seas of the contemporary era, since the modern sea levels were achieved by around 3000–4000 years B.C. Clearly Howieson's Poort represented the coldest seas or the most glacial conditions, whereas so-called MSA II (Level 15, Cave 1) is still considerably cooler than modern conditions. but less cold than those of Howieson's Poort. So far, this fits with the stratigraphic interpretations, and places the Howieson's Poort of Shelter 1A as roughly contemporary with the depressed sea levels of Level 13 in Cave 1. Level 15 (MSA III) then represents higher-water levels but still clearly glacial conditions relative to the modern situation.

The chronological difficulty comes when we look at the isotope data from the so-called MSA I. The single shell from Level 38 of Cave 1 and the three shells from KRM-5 are similar, being just slightly different from the mean for the LSA shells that were taken as controls indicative of modern interglacial conditions. When Butzer saw this situation, he reasoned that the only time in the recent Pleistocene past when interglacial conditions has obtained was during the Eem (or Riss–Würm interglacial of the French sequence) dating around 125,000 years B.P. It is the facts of a similarity between the oxygen isotope values for the two shells from LSA levels and the four shells from the alleged MSA I levels that provide the basis for the inferred chronology of the MSA. Although it is true that Shackleton included the shells from Shelter 1B in his sample of MSA I, I find that the latter has significantly higher isotopic values than are seen for the Cave 1 and Shelter 5 samples. This fact lends support to my earlier skepticism regarding the assumption by Singer and Wymer (1982) that Shelter 1B was contemporary with the MSA deposits in Cave 1. I consider the isotopic data support for a temporal placement for Shelter 1B deposits either in late MSA II times, or alternatively after Howieson's Poort. Leaving this detail aside, the current chronology for the MSA rests on the accuracy of $^{16}O/^{18}O$ ratio readings on six shells, two from LSA, and four from the MSA I. The constructive skepticism I advocated in Chapter 1 leads me to be very uneasy with this chronology, and to be even more uneasy with its derivatives (Beaumont 1980; Klein 1980).

Radiocarbon Dates

A total of 33 ^{14}C dates were obtained on samples from Klasies River Mouth levels. These were all processed by Geochron Laboratories. The most that can be said of these dates is that 18 of the 27 dates (exclusive of the 6 dates on the LSA) were maximum dates; that is, the ^{14}C was not present at measurable levels, suggesting ages for the specimens greater than the elapsed time level indicated. Stated another way, the vast majority of the specimens analyzed by Geochron were beyond the range of ^{14}C dating techniques. Three additional dating tests were run by the South African National Physical Research Laboratory at Pretoria on specimens also analyzed by Geochron. Two of the three specimens had yielded finite dates according to Geochron analysis; yet the Pretoria lab reported maximum dates. This all strongly suggests that the entire MSA sequence at Klasies River Mouth, including the MSA III levels, dates beyond the measuring capabilities of the ^{14}C method—that is, before 40,000 B.P. This opinion supports the earlier arguments of Beaumont and Vogel (1972) that the short chronology for the MSA is suspect. It does not, however, support the particular inferential chronology that has been generated almost as a series of conventions since it was originally proposed by Butzer; that is, the assignment of MSA levels to various stages of the oxygen-isotope chronology for ocean temperatures.

Summary

It is apparent to me that the old pre-^{14}C ideas of southern Africa chronology are obsolete. Reliable dating has pushed the LSA back to around 20,000 B.P. Some sequences, such as Klasies River Mouth and perhaps Border Cave (Beaumont 1980), suggest that much of the MSA represented at least in those deposits is older than the ^{14}C methods are able to measure reliably.

Because of the absence of a reliable dating method, in South African archaeological circles the convention arose of assigning levels from key archaeological sites to stages and substages of the Pleistocene climatic sequence, as known by oxygen-isotopic evaluations of deep-sea cores extracted from the ocean bottoms. There is no basis for this method that I can see except rather nonspecific ideas about what terrestrial environments should have been like during cold versus warm stages (see Klein 1980). In all cases of archaeological sequences treated in this manner, the sites lack pollen records, so the climatic interpretations are made on sedimentological grounds and/or on the basis of inferences from the faunal remains (see particularly Beaumont *et al.* 1978; Klein 1980; Volman 1981).

In my opinion these schematic interpretations are essentially a waste of time because (1) we do not know the character of the processes determining climatic change, and (2) the assumption that variability in faunal remains is a simple reflection of the population of animals available in the gross environment, unfiltered by the techniques of procurement and the character of the niches occupied by the hominids, is tenuous at best. We should be seeking instead to understand the character of hominid behavior.

> In East Africa, animal bones recovered from the world's oldest archaeological sites, dating to between 2 and 1.5 million years ago, tell us that by then meat was a regular element in the diet (and also that people had acquired the typically human habit of leaving food waste and other garbage at a repeatedly occupied "home base"). (Klein 1979:151)

To consider all the bones accumulated in archaeological sites as there due to hominid hunting is to assume what we seek to understand. To assume further that hominid hunting took essentially a random sample of the animals available in the habitat is seemingly unjustified at any level. And then to assume that the bones acquired by archaeologists were present by virtue of the past operation of that "typically human habit" of leaving food waste and other garbage at repeatedly occupied home bases is again to assume the very condition of land use and behavior that we seek to investigate. The currently prevalent MSA chronology in southern Africa is based, unfortunately, on these "cart-before-the-horse" types of assumptions.

As has been pointed out, this study intends to develop methods for understanding hominid subsistence practices. It will be interesting later to consider the degree to which any gains in this direction affect the almost purely climatic interpretations of species frequencies (see Klein 1980:253) that currently dominate the attempts among Africanists to use fauna as a dating medium.

I have tried to characterize the situation as archaeologically investigated at Klasies River Mouth. I have treated the interpretations developed by the primary workers most intimately involved. My criticisms have been largely limited to the problem of the cross-correlation of the levels of Shelter 1B and to the confusing or contradictory arguments regarding environmental correlates used as a basis for dating the deposits. The dating as summarized here has been widely generalized and used as a basis for cross-correlation with a number of other sites in the southern African areas (see Beaumont *et al.* 1978; Klein 1980; Volman 1981). These cross-correlations—both among the sites at Klasies River Mouth and among sites from other regions—have served as the basis for claims that fully modern man appears earlier in southern Africa then in any other place thus far known (see Beaumont *et al.* 1978; Rightmire 1979), and that Howieson's Poort industry is a precursor of the Upper Paleolithic, consistent with the early appearance of fully modern humans in the area. If these claims are accurate, and if at the same time we can reasonably suspect a significant degree of scavenging on the part of the Klasies River Mouth occupants, then this site truly has revolutionary implications for our current views of the past.

CHAPTER 3

The Klasies Fauna: Approaches to Analysis

Introduction

I have given the reasons why I chose to study the Klasies fauna. I have discussed something of the character of the site and the previous work that has been directed toward dating the deposits and relating the materials to other MSA data from southern Africa. Now I must deal with some of the characteristics of the fauna itself because the integrity of the deposits must be assessed before the contents can be related to hominid behavior. In other words, I must be able to assess the extent to which observations made on the fauna are relevant to hominid behavior, and not to the behavior of other bone-accumulating animals. In addition, I need to clarify my procedures for observation and recording of the fauna.

The assessment of the integrity of the deposit can only be made after the fauna has been observed and tabulated—which means that it is essentially a conclusion about the fauna. Nevertheless, the logic of presentation demands that the reader be assured which agent the facts refer to. That is, we need to clarify the integrity of the faunal assemblage before proceeding to the description and analysis of facts thought to inform about hominid behavior.

Approaches to Description

Richard Klein (1976) has previously described the fauna from this site. Klein had sorted the bones into boxes identified by species and anatomical parts for the site as a whole. I examined each box in turn, trying to remain as faithful to Klein's earlier study as possible. I should point out, however, there is a point of disagreement—or perhaps it is a more tactical difference—in the ways we studied this fauna. Klein tabulated what he calls *MNI*, or minimum number of individuals. He considers age estimates of animals at time of death (whether bones exhibit different patterns of epiphyseal union, or whether teeth exhibit different patterns of crown wear) when tabulating whether one or more individuals are represented by homologous anatomical parts. Klein also reports the number represented by the maximum number of parts from the right or left side of an animal. That is, if there are four left calcanea and six right ones, he reports a minimum number of six individuals. (Klein is roughly following the procedures as outlined by Chaplin [1971] for reporting minimum numbers of individuals.)

I have strongly criticized this procedure (Binford 1978:67–72, 478–479; 1981; 1982) because, in my experience, the bones recovered from human sites do not represent complete individuals that were once there (as MNI methods assume), but instead represent the biased introduction of already selected and sometimes processed anatomical segments to a site either for further processing and/or storage, or for consumption. This means that man does not kill an animal, begin eating at the tail and eat through to the head, and then go off in search of another animal. Quite to the contrary, he kills an animal and, depending upon the size and condition of the prey relative to the food demands among the consumer population the hunter is serving, he butchers the prey differentially, frequently disposing of some parts judged to be of marginal utility because their transport is troublesome. Or, if the distance is great, he may process some parts so as to reduce their bulk. For instance, the hunter may fillet some of the meat from a ham area, and transport only the *biltong*, or partially dried meat strips, leaving behind the bones of the rear leg. On the other hand, the hunter may also make very different decisions depending on the size and condition of the animals killed. For example, the Nunamiut Eskimo consider the fatty morsels of tissue within the skull to be highly desirable fare if the animal is young and tender. This means that in any living site to which parts are carried with some difficulty, as when accessible only by walking, it is almost a certainty that the heads introduced are from young calves or juvenile cows. The heads of adult cows and bulls would be either eaten by the hunters at the kill site or abandoned unprocessed in the field. Thus the quality of the food versus the difference in size of the part between calves and big bulls ensures that the same part of animals of different age will be treated differentially. On the other hand, the meat of the shoulder from large, mature animals will be the part transported, whereas the shoulders of the young animals are considered to be too stringy to warrant carrying so many bones for just a little bit of bad lean meat.

In short, man does not normally consume meat in anatomically complete units. He transports and differentially uses anatomical segments dismembered from whole animals. The presence of a particular segment at a site does not imply that the entire animal, anatomically speaking, was ever there. But the assumption of equivalence between a part and the whole animal is what stands behind attempts to infer (1) the total meat available to a group of hunters from the MNI represented by given anatomical parts; or, even more importantly, (2) the constitution by age and sex of the population of animals as killed from the biased distribution of anatomical parts from animals of different sex or age.

Richard Klein has sought over the years to develop ways of using properties of teeth for identifying the age profile of the animals represented by teeth at archaeological sites, with the clear suggestion that the patterning recognized was referable to bias in the population of animals killed or dying as a function of the hunting strategies of the human group (see Klein 1982). I suggest that Klein's approach is an extension of the MNI tactics that seek to equate numbers of anatomical parts to numbers of animals killed or available for consumption at time of death. My approach takes numbers of anatomical parts at face value, fully recognizing that man may well select or process anatomical segments differently. This means that we may see one age or sex bias present in one anatomical part, but perhaps a different bias associated with another part. This situation is a textbook example of the failure of Klein and others to investigate the formation processes (see Schiffer 1972) of the archaeological record. They have instead assumed a direct relationship between age or sex bias in archaeological remains as referring unambiguously to age or sex biases in the death population of the animals represented. Similarly, frequencies of a given anatomical part from an animal have been accepted as representing the prior presence at the site of the complete animal.

My studies of hunter–gatherers (Binford 1978) and nonhuman predators (Binford 1981) clearly illustrate that these assumptions are not justifiable as general conditions. Such direct equations may not be taken for granted when making inferences from archaeological observations. Many other selective decisions and situations affecting the frequency and/or visibility of anatomical parts stand between the archaeological record as observed and those animals that, in their death, served as sources for the parts being studied. For these reasons, I have decided to reduce the ambiguity of language by no longer referring to anatomical frequency counts as MNIs. Throughout this report I speak instead of *MNE* (minimum number of elements) and *MAU* (minimum animal units).

Minimum numbers of elements are just that—the minimum number of different specimens referable to a given anatomical part used in classification. For instance, the proximal humerus may be the class, and I may seek to tabulate the numbers of this unit, yet the proximal humeri present may all be broken. My task is then to estimate the minimum number of proximal humeri represented by the fragmentary remains, on the assumption that a complete or anatomically recognizable proximal humerus had originally been present. In this procedure, the inference between the population of units observed archaeologically and the population that once existed in the past is made at the anatomical segment level rather than at the level of whole animal units. Given my focus on anatomical segments, I ignore differences between sex, age, and side, since these are properties of animals. A proximal humerus is still a proximal humerus, regardless of left-versus-right orientation on the animal, or the age or sex of the animal from which it was derived. I am estimating minimum numbers for the anatomical category designated; for example proximal humeri. This is not to say that age, sex, and side information are not useful, or should not be regularly recorded in studying fauna, only that these are not attributes that define minimum numbers of anatomical elements.

Minimum animal unit is a conversion for comparative purposes in which the anatomy of a living animal is taken as the standard for reporting frequencies of minimal numbers of elements. Since the living animal has two proximal humeri, the MNE count is divided by two so that the frequency of the element is expressed in animal units for comparative purposes. I am not suggesting that MAU in any way implies the number of animals that were actually killed to account for the fauna studied. My interest is not in how much meat was available to an ancient group, or even in their hunting biases. It is rather in their procurement, transport, processing, and consumption tactics when using animal products.

Given such interest, it is important to be quite confident that the bones studied were introduced and accumulated as a consequence of hominid behavior. It must be recognized that many sites that yield the unambiguous products of man also contain products of other agents, particularly nonhominid carnivores and scavengers, as well as other nonhominid inhabitants of caves and protected shelters. I have previously (Binford 1981) pointed out the fallacy of accepting the total inventory of bones recovered from a site that yields stone tools as necessarily referable exclusively to the behavior of hominids.

Now, how do we assess the taphonomic integrity of the Klasies materials?

Evidence for Use of the Site by Porcupines

Approximately 3% of the faunal elements counted in my study showed evidence of having been gnawed by porcupines. This evidence is neither

Figure 3.1 Porcupine gnawing on the proximal end of an eland (*Taurotragus*) metacarpal (Cave 1, Level 14).

subtle nor difficult to evaluate. Figure 3.1 illustrates a classic example of porcupine gnawing. Table 3.1 summarizes the number and percentage of MNEs (as given in Table 3.5) identified in my basic list of anatomical parts, separated into the same body-size class in terms of which the fauna from the site was studied. Opposite the "number" column for each sample is the percentage of the total elements that were gnawed by porcupine.

Several facts are of interest here. None of the bones from the small animals were gnawed by porcupines, and only two examples were noted on bones from body-size class II (for body-size classes, see pp. 77–78). Beginning with bones from body-size class III, the frequency of gnawed bones generally increases with body size for biased group of anatomical elements. A control body of data has been described by Brain (1981:302) for collections made from a porcupine lair in the Nossob River area. There, the exclusive agent of bone transport to the site was the porcupine. There was an analogous near-avoidance of bones from very small animals, whereas bovid size-class II, unlike the situation at Klasies River Mouth, was moderately well represented and size class III was most common. Unfortunately in the Nossob case, animals of body-size classes IV and V were not represented, and they were not believed to be generally present in the habitat.

TABLE 3.1

Porcupine Gnawing and Transport Comparisons

Anatomical part	Control data: large-prey kill sites[a] (1)		Nossob porcupine lair (1956)[b] (2)		Nossob porcupine lair (1968)[c] (3)		Klasies Cave 1 porcupine-gnawed bone (4)	
	Number	*%*	*Number*	*%*	*Number*	*%*	*Number*	*%*
Head	165	.19	195	.31	62	.19	23	.28
Vertebrae and pelvis	486	.34[d]	182	.29[d]	148	.46[d]	10	.12
Ribs	46	.03	38	.06	5	.02	0	.00
Upper-front leg	247	.17	66	.10	40	.13	9	.11
Upper-rear leg	183	.13	40	.06	39	.12	1	.01
Lower-front leg	70	.05	19	.03	4	.01	10	.12[d]
Lower-rear leg	160	.11	50	.08	10	.03	26	.31[d]
Phalanges	66	.05	45	.07	12	.04	4	.05
	1423		635		320		83	

[a] Data taken from Binford (1982:Table 5.08, Column 2). The index percentages were simply added up as a guide to frequencies. The reader may wish to compare this to Behrensmeyer and Dechant Boaz (1980:85).

[b] Data taken from Brain (1981:Table 60), omitting sesamoids and metapodial pieces.

[c] Data taken from Brain (1981:Table 60, p. 304), omitting sesamoids and metapodial pieces.

[d] Underlining is for emphasis.

I have collapsed the frequencies of parts tabulated from the Nossob lair into gross anatomical classes (Table 3.1). I have also collapsed the standard faunal list that I generally use for the control population of faunal remains collected from kill sites, or carcasses remaining on the land surface after predators and scavengers depart. These are all compared to the same anatomical classes of gnawed bones observed at Klasies River Mouth.

It is very clear that the bones introduced to the Nossob lair appear to have had the same composition as a population of bones remaining on the landscape after predators and scavengers had finished (Column 1). This is made even more clear when both my control data and the Nossob lair data are compared to the similar percentages for control surface collections of bones made by Behrensmeyer and Dechant Boaz (1980:85). Stated another way, the porcupines are essentially randomly sampling the population of bones available to them for exploitation. This point was made even more explicit by Brain's demonstration that the bones in the Nossob lair from different species were nearly identical to the proportions of the animals actually present in the habitat (see Brain 1981:114–115).

Given this type of behavior—the random transport and/or gnawing of bones available to the porcupines—we would expect variations in the content of anatomical parts to reflect directly the composition of the population of bones from which the porcupines drew their samples. Looking at Table 3.1, Column 4, we immediately note that the percentages of porcupine-gnawed bones at Klasies River Mouth are considerably different from those either in the Nossob lair, or in the control population of bones apt to be available to a porcupine for transport. There is a striking inflation of the parts of the lower-front and -rear legs that have been gnawed. There is an equally interesting, abnormally low value for vertebrae and pelves. Stated another way, the porcupine-gnawed-bone frequencies at Klasies River Mouth yield a head-and-lower-limbs-dominated graph, whereas for the Nossob lair and the control data a head-plus-axial-skeleton-dominated graph is noted. Given that porcupines sample the bone population available to them, it seems clear to me that the gnawed bones in Klasies River Mouth are not bones introduced to the site by porcupines, but instead are bones selected from the site for gnawing. After all, the site was selected for study because of its characteristic head-and-lower-limbs anatomical-part profile.

Clearly, porcupines inhabited the site, but they seem simply to have taken advantage of the large number of bones already introduced there. This is further evidenced by the bone in Figure 3.2, on which the porcupine tooth marks pass over marks inflicted on the bone by stone tools. Three such examples were noted in the assemblage of porcupine-gnawed bones from Klasies River Mouth.

It should be pointed out that of the 83 porcupine-gnawed bones from

Figure 3.2 Porcupine gnawing over a tool-inflicted mark on a vertebral fragment (Cave 1, Level 14).

the Cave 1 deposits, all except 8 were from Level 14. Of these 8, 3 were from Level 17b, 3 from Level 16, and 2 from Level 15. Clearly, the most significant porcupine occupation was coincident with the accumulation of Level 14, which is, as previously pointed out, largely a secondary deposit. It may well have accumulated when considerable erosion of the scree deposits outside Shelter 1A was occurring. All these conditions suggest that man's presence may have been minimal during the accumulation of this deposit.

I conclude that porcupines did in fact inhabit Klasies Cave 1, primarily during the accumulation of Level 14. They do not seem to have biased the faunal population by their bone-collecting and -gnawing activities. It is evident that they collected their bones on site from those already introduced by other agents, and did not substantially modify the character of the faunal assemblage recovered from the site.

Evidence for Leopards

The excavators report that a nearly complete leopard skeleton was recovered on the interface between Levels 14 and 15. Klein (1976:91) lists leopard bones of all major anatomical classes from Level 14. He estimates that there was a minimum of four individuals represented in the level. This is consistent with the observations of others (Brain 1981:82–89) that leopards are commonly represented in faunal accumulations where their feeding behavior is a common source for the deposit. Leopards have been noted to bring prey parts back to breeding dens and into protected, sheltered locations where they may spend many of the daytime hours. In such cases, the lair may be used for many years, yet the actual numbers of accumulated bones may not be impressively large (see Brain 1981:88). What function the caves and shelters at Klasies River Mouth may have played for leopards is not clear, but the number represented in Level 14 strongly suggests that leopards may well have contributed to the faunal assemblages, particularly in that level. While Level 14 has a high species count, there are more baboons represented there than at any other level in the site. It has been well established that leopards do systematically take baboons, particularly in their sleeping places at night (see Brain 1981:84). It would thus appear that the integrity of the contents of Level 14 is suspect as regards relative species frequencies.

As a check on the degree that Level 14 differs from the other levels in the site in relative anatomical-part frequencies, I prepared Table 3.2 and Figure 3.3 from R. Klein's (1976: Tables 9–13) original level-by-level inventories. The animals in body-size classes IV and V are generally beyond the size range of normal leopard prey (see Brain 1982: Figure 97), so it is highly unlikely that leopard occupation would affect these faunal frequencies. The bones from Level 14 are compared to the sum of the inventories fro all the other Cave 1 levels.

The characteristic head-and-lower-limb pattern of part presence is well illustrated (Figure 3.3). However, this pattern is more robust in the Level 14 sample than is the case for the combined sample from the other Cave 1 samples. This means that there is nothing unique about the Level 14 fauna that would differentiate it from the remainder of the site.

Turning now to the comparative graphs (Figure 3.4), in the two smallest body-size classes of bovids—the size range in which most leopard prey would be expected—we note the pattern of parts represented in Level 14 is not very different from that seen for the remainder of the site. The major difference seems to be that, in general, the bones of the smaller animals are less well represented in Level 14 than they are in the other levels. That is, the graph of Level 14 is generally lower for most of the vertebrae, metapodials, and (importantly) elements of the rear leg, whereas parts having high intrinsic survival potential seem inflated in their numbers. This contrast strongly suggests that the major difference between Level 14 and other levels in Cave 1 may be some sorting and/or differential operation of an attritional agent

TABLE 3.2

Comparison between Level 14 and Other Cave 1 Levels for Recognizing Bias That Might Result from Leopard Use of the Site[a]

	Size class									
	IV and V				*I and II*					
	Level 14		*Other levels*		*Level 14*		*Other levels*		*Leopard lairs*	
Anatomical part	*MNE (1)*	*% (2)*	*MNE (3)*	*% (4)*	*MNE (5)*	*% (6)*	*MNE (7)*	*% (8)*	*MNE (9)*	*% (10)*
Maxilla	29	.69	85	.63	17	.50	26	.39	21	1.00
Mandible	42	1.00	134	1.00	26	.76	55	.83	2.5	.63
Atlas	8	.19	6	.04	7	.21	8	.12	1.0	.25
Axis	5	.12	10	.07	5	.15	18	.27	1.0	.25
Cervical vertebrae	7	.17	11	.08	4	.12	13	.20	1.8	.45
Thoracic vertebrae	3	.07	8	.06	3	.09	12	.18	1.3	.33
Lumbar vertebrae	4	.10	8	.06	4	.12	14	.21	.75	.19
Pelvis	9	.21	15	.11	13	.38	38	.58	1.0	.25
Scapula	8	.19	10	.07	34	1.00	66	1.00	1.5	.38
Proximal humerus	4	.10	3	.02	3	.09	7	.11	1.0	.25
Distal humerus	13	.31	15	.11	12	.35	13	.20	0.5	.13
Proximal radiocubitus	20	.48	20	.15	6	.18	9	.14	1.5	.38
Distal radiocubitus	10	.24	9	.07	4	.12	8	.12	1.0	.25
Proximal metacarpal	18	.43	57	.43	1	.03	7	.11	1.5	.38
Distal metacarpal	11	.26	20	.15	1	.03	8	.12	1.5	.38
Proximal femur	8	.19	10	.07	7	.21	16	.24	1.5	.38
Distal femur	4	.10	11	.08	6	.18	19	.29	1.5	.38
Proximal tibia	2	.05	2	.01	4	.12	15	.23	2.0	.50
Distal tibia	15	.36	22	.16	5	.15	22	.33	2.5	.63
Tarsals	20	.48	49	.37	—	—	—	—	1.0	.25
Astragalus	36	.75	28	.21	5	.15	16	.24	2.0	.50
Calcaneus	16	.38	24	.18	8	.24	28	.42	1.5	.30
Proximal metatarsal	21	.50	31	.23	4	.12	10	.15	1.5	.38
Distal metatarsal	6	.14	10	.07	3	.09	13	.20	1.5	.38
Phalanges	10	.24	28	.21	2	.06	11	.17	1.75	.44

[a] Data for Columns 1, 3, 5, and 7 taken from Klein (1976:Tables 9–13). Data for Column 9 were synthesized from Brain (1982:Figure 90, and "Klipspringer and Class II antelope" of Tables 42, 43, and 44).

that deletes small-animal parts in a biased manner. This could be carnivore feeding, but I suspect that is more likely geologic sorting, which is certainly expected in Level 14, because it has been interpreted as, at least in part, a secondary deposit.

Compared to the graphs from Level 14 is a graph of small-bovid parts

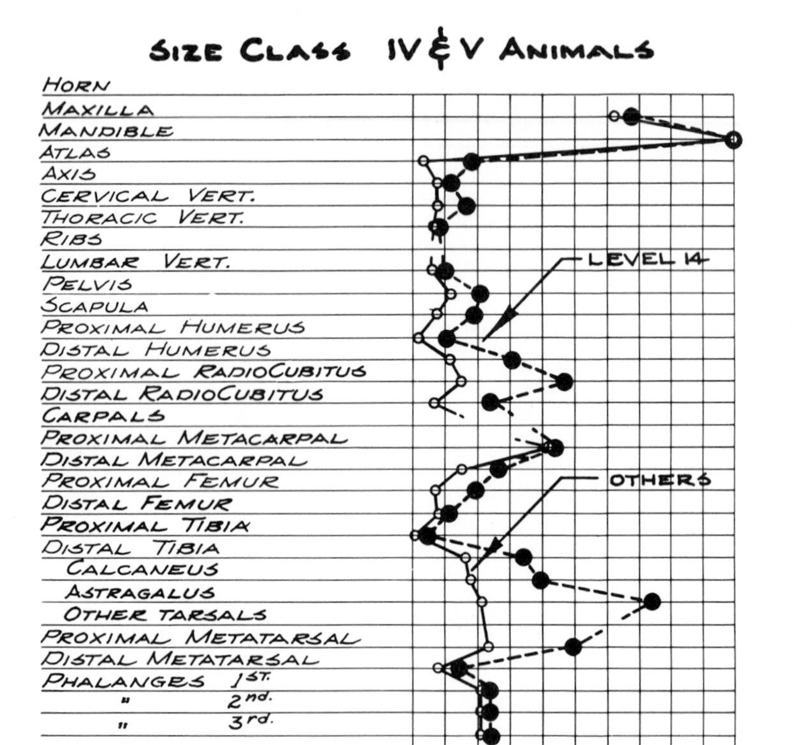

Figure 3.3 Graphic comparison between large-bovid remains from Level 14 and the combined frequencies from all other levels at Cave 1.

that were recovered from three separate leopard lairs as reported by Brain (1981 Figures 90 and 91; pp. 296–297). It is clear that the part bias favoring cranial parts at the expense of scapula and pelvic parts (which dominate the Klasies data) does not appear manifest in the contrast between the Level 14 small bovids. From this I conclude that there is no recognizable influence on the small-bovid faunal population from Klasies Cave 1, Level 14, referable to leopard feeding. The presence of leopards, however, certainly suggests that this should be expected. Its apparent absence may rest with either the failure to excavate or the poor geologic preservation of the area of the cave where leopard feeding was most common, since it is observed that the faunal results of leopard meals are not randomly scattered throughout a lair (Brain 1981:89) It has also been noted that leopards introduce much less prey into breeding lairs (Brain 1981:144). Consistent with this suggestion Klasies may have been used in this manner so that at least some of the leopard remains may derive from the actions of hominids, perhaps in securing their sleeping areas. In any event I see no reason to question the integrity

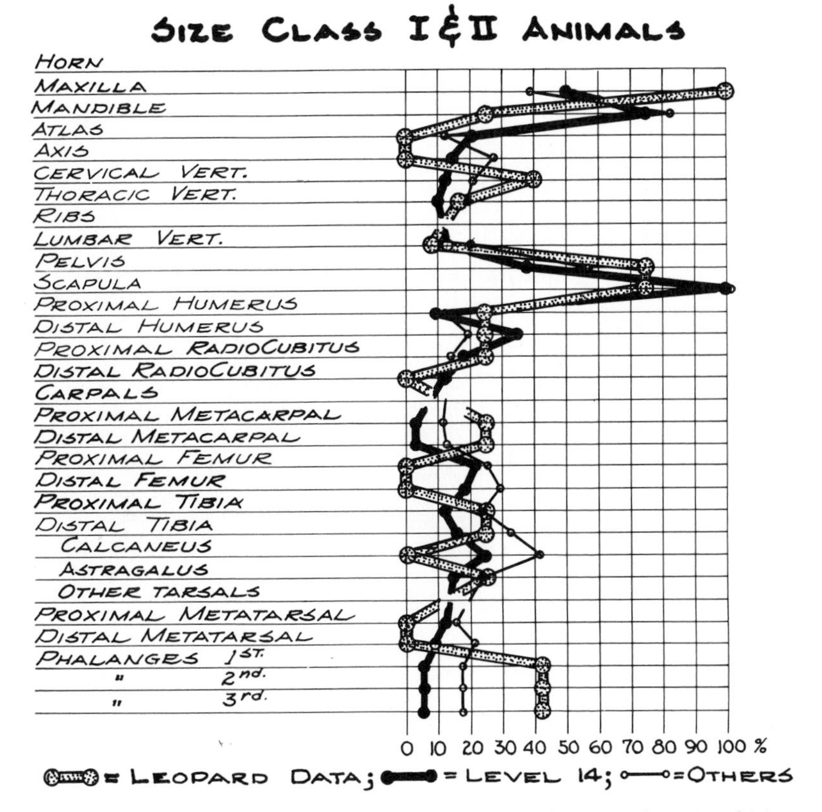

Figure 3.4 Graphic comparison between small-bovid remains from Level 14 versus the combined "other" levels at Cave 1, and the control data from leopard lairs collected by C. K. Brain.

of the Klasies deposits relative to leopards since leopard feeding seems to be not at all in evidence.

Evidence for Hyaenas

I have seen no evidence for the role of the hyaena in the accumulation of the Klasies River Mouth fauna. One would expect that any accumulation on site would be in the context of hyaena denning. I have had the opportunity of observing eight hyaena dens, six occupied by brown hyaena (the form most likely to have been present near Klasies River Mouth) and two dens of the spotted hyaena. All of these were excavations into sand, generally at the base of a tree so that the roots tended to support the roof and

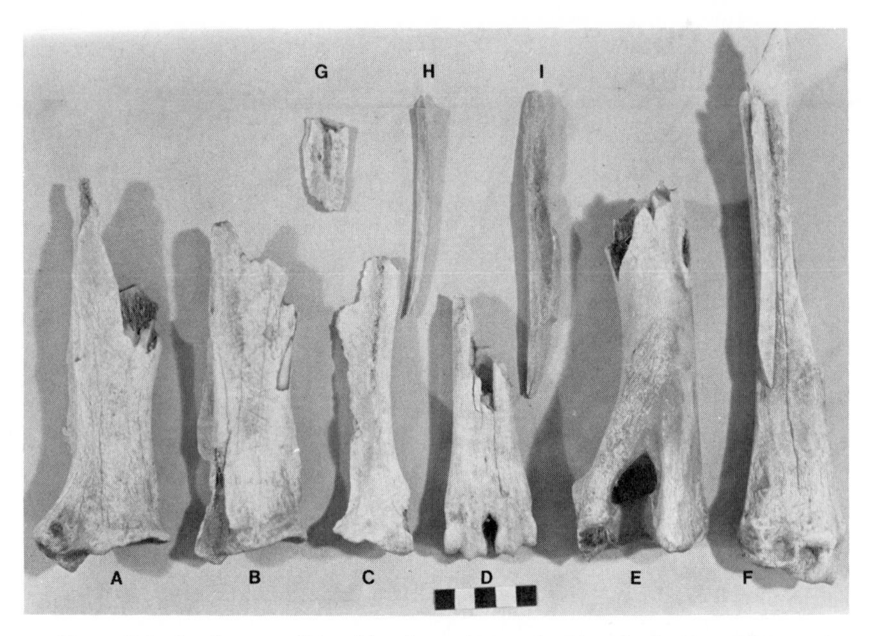

Figure 3.5 Leg bones collected by the author at the Grootbrak spotted hyaena den in the Nossob River valley, showing typical patterns of breakage and gnawing.

render the den usable for a substantial period of time. Nevertheless, these excavated dens are not occupied for really long periods of time since they do collapse. The result is that large assemblages of bones do not accumulate around them as is the case when rock fissures and caves have been used. Nevertheless, I was able to collect a moderate sample of bones from around dens, and several things are of relevance for assessing the Klasies fauna. As I reported previously (Binford 1981) for wolf dens, gnawed bones are common on den sites, particularly for the spotted hyaena, and these may be gnawed extensively (see Figure 3.5 and the detail of the tooth scoring shown in Figure 3.6). This type of extensive gnawing is absent on the bones at Klasies. I report in a later chapter the gnawed bones from Klasies. These are most often bones with tooth punctures in soft cancellous tissue or mashed and "scooped-out" areas of the soft articular ends. The occasional tooth scoring noted around articular ends is generally restricted to a few parallel marks, transverse to the longitudinal axis of the bone. The extensive mouthing of bones seen in the hyaena dens is not in evidence at Klasies River Mouth.

The mouthing of bones by the hyaena produces beautiful pseudotools of forms not seen among wolves and other canids, such as coyote, fox, and dog. Figure 3.5 shows the disarticulated limb-bones collected at one spotted hyaena den (Grootbrak Den). Figure 3.6 is a detail of the tooth scoring on

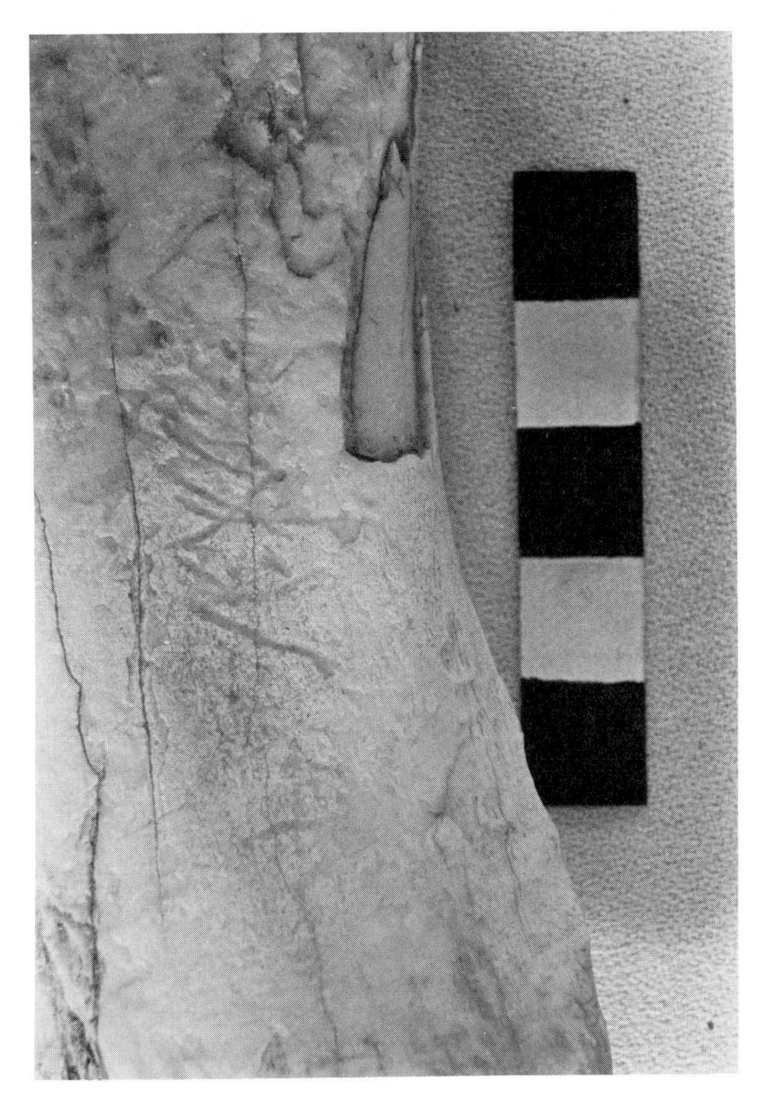

Figure 3.6 Detail of tooth scoring on a scapula shown in Figure 3.5B.

the scapula in Figure 3.5B and Figure 3.7 is the smoothed and worn point of broken proximal end of the radius shown in Figure 3.5F. This type of mouthing of bones that produce highly polished and chipped edges was not observed at Klasies River Mouth. The type of animal gnawing noted on the Klasies bones was consistently more akin to that illustrated in Figure 3.8, which shows a canine puncture mark and mashed and scooped-out cancellous tissue. (For more information, see Maguire and Pemberton 1980).

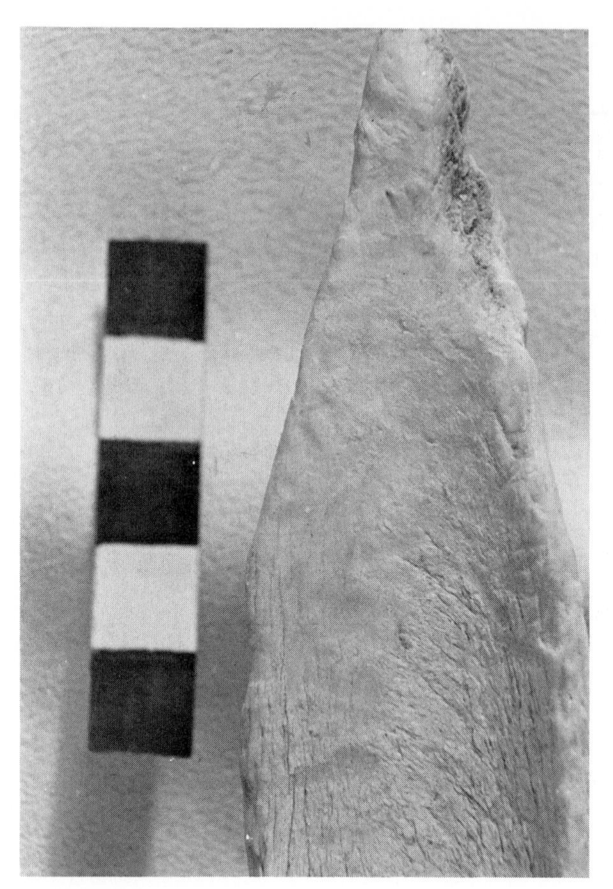

Figure 3.7 Detail (from Figure 3.5F) of chipped and smoothed pseudotool produced by spotted hyaena.

Perhaps the most telltale characteristic of hyaena accumulations are (1) a fair proportion of complete bones; (2) many bones with one articular end and a substantial section of attached shaft; (3) a large number of bone-shaft cylinders, as illustrated in Binford (1981:173; 1982a:178); and (4) very few bone splinters and segments of diaphysis.

Although bone splinters were very common at Klasies, not one bone cylinder was observed. Complete bones were rare and then restricted to the smallest bovid species (a situation opposite to that in hyaena accumulations) and most long bones were broken near the articular ends with impact fractures, characteristically made when a bone is broken by percussion rather than gnawed down attritionally the way animals reduce bones (see Binford [1981:169–181] for a description).

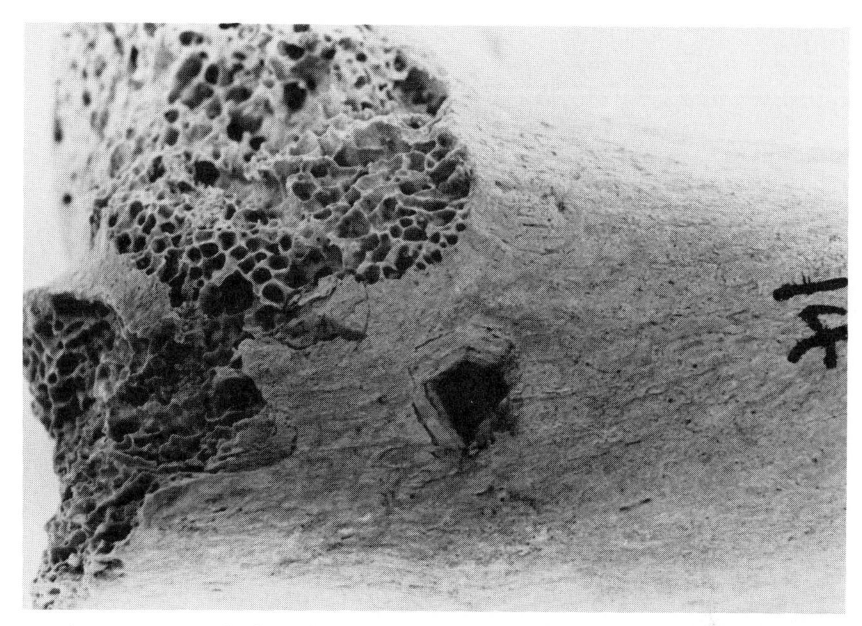

Figure 3.8 Detail of tooth puncture and scooped-out soft tissue on an articular end—example of typical modifications on bones from Klasies River Mouth excavations.

Finally, the anatomical-part frequencies at Klasies River are completely out of line with the parts most commonly observed in hyaena dens. In my sample, horn and skull fragments coupled with neck parts were most common (for additional data see Skinner *et al.* [1980]). However, lower limbs may dominate in some cases, but these are frequently less fragmentary than upper-limb bones. As we will see, the *most* intentionally and extensively percussion-fractured bones Klasies River Mouth are those of the lower limbs from medium- to large-body-size animals. In my opinion, the case for hyaena as an agent of bone accumulation at Klasies River Mouth simply cannot be made.

Assessment of Integrity

Clearly, African porcupines were active in Klasies Cave 1, particularly during the period of accumulation of Level 14. They chewed on bones, leaving their telltale tooth-inflicted marks. They do not, however, appear to have introduced bones to the deposit from sources other than the surrounding archaeological deposits. The comparisons with other known porcupine accumulations clearly indicate a biased chewing of parts consistent with the anatomical-part-biased archaeological deposits that yield no evidence of porcupines. In short, they were gnawing on bones already at the site and were not modifying the population through the regular introduction of parts from nonarchaeological sources around the site.

The evidence for leopards is provocative, particularly in the presence of leopard bones themselves. Interestingly, like the porcupine evidence, this is concentrated in Level 14. Nevertheless, the population of small-bovid parts in Level 14 is consistent with the part frequencies from levels not yielding leopard parts, and both are different from the inventory of parts thus far documented for small bovids from leopard lairs. The activity of leopards is suspected but does not appear to have impacted the faunal population excavated from remnant deposits in Level 14.

Finally, the evidence for hyaena use of the site as a den is nil. The result of this assessment is that the faunal contents of the levels as excavated may be taken as predominantly the result of hominid behavior, and any influence of other bone-accumulating agents appears to be minimal to nonexistent, and, if present, largely restricted to Level 14.

Units of Observation

The archaeologist observes an archaeological deposit in terms of some unit, the contents of which are considered for analytic purposes to be potentially meaningful. In general, excavators seek "natural" units, such as geomorphological components or inferred depositional strata as basic units of observation. In addition, many excavators use arbitrary reference units, such as squares or levels, in terms of which records of placement for the contents of deposits are generally kept. This was true during the excavation of Klasies River Mouth, and in the initial analyses by the excavators (Singer and Wymer 1982) and Richard Klein (1976). Klein's publication (1976) reports the frequencies of anatomical parts from different body sizes of animals by level, as generally described in the discussions of the site. Klein, however, generalized from these observations a pattern; namely, that small animals were represented by more complete anatomical inventories than were the larger animals. Further, the larger the animal, the greater the probability that it was represented almost exclusively by head and lower legs. These generalizations, which attracted me to the study of Klasies material, were offered as characteristic of all the units of observation from Klasies River Mouth. Of course, it is possible that the patterning described by Klein is the chance result of collapsing a number of unlike subpopulations into giant palimpsests, the combined properties of which are different from the properties of any of its contributors.

Actually, even a casual inspection of the tables presented by Klein (1976:Tables 9–13) substantiates his claim. The level-by-level tabulations show little variability in anatomical-part frequencies, and all are basically consistent with Klein's generalizations. This is the type of pattern that we as archaeologists should be seeking to recognize (see my criticism of the episodal reconstructionist's view in Binford 1981:197–198). A major pattern such as that suggested by Klein's comparative work demands explanation. If we can understand such a macroscale pattern, then we are taking a major step forward in our ability to understand something of the character of the past. Because I sought to understand the processes that operated to create the patterining described by Klein, I chose to use as my initial unit of observation the entire population of bones recovered from the MSA levels of Klasies River Mouth Cave 1. My interest was to see if there were not some additional properties of the bones that could be observed to vary in a manner patterned at the same scale as generalized by Klein.

Klein had already sorted the fauna from Klasies River Mouth into boxes of anatomical parts segregated by species. I studied these units, species differentiated by body part. My observations, then, were in terms of the analytic units used by Klein for describing the assemblage composition of the original excavators' units of observation, the levels and squares. The species units were, of course, the units in terms of which Klein recognized the provocative differential distribution of anatomical parts among species of different body size. My procedure was to tabulate the frequency of MNE for each species and to make observations on the frequencies of inflicted marks, breakage, and other clues to the past history of the bones by species and anatomical-part classes. If I could understand some of the patterned variability noted within the faunal population in these terms, and if that variability was correlated with the anatomical-part frequency patterning generalized by Klein, then I might have a foot in the door to evaluate the meanings to be attached to the patterning described by Klein.

This point brings me full circle back to a basic issue raised in my introduction—namely, how do we use our available knowledge to guide our observations?

Information Guiding Observation and Analysis

As I have suggested, I used knowledge regarding the parts apt to be exploited by hunters versus those that are apt to be available to a scavenger as a guide to selecting the Klasies River Mouth fauna as a target for study.

Given the opportunity to study the fauna, I pursued certain systematic lines of inquiry and recording, particularly with regard to (1) breakage of anatomical parts, (2) inflicted marks from tools, (3) inflicted marks from animal teeth, and (4) evidence of burning. These classes of phenomena were chosen because I already had some experience that led me to believe that in patterned combinations of these properties, coupled with anatomical-part frequencies, there might well rest criteria unambiguously diagnostic of scavenging.

When developing ways of recognizing scavenging through the study of animal bones, we must put into place the *controls*—the known conditions that we use in establishing diagnostic characteristics indicative of the actions in the past we seek to know. Of fundamental importance is the recognition that when a human—or any agent, for that matter—selects and removes anatomical parts from a larger population, the proportions present at the time of selection and the condition of the parts available for selection influence the numerical form of the population selected, as well as the formal condition of the parts in that population as seen at a later time. In less-abstract terms: When a hunter kills an animal, the entire complement of anatomical part is available for selection and use by the hunter. In addition, all the bones are in "mint" conditions; that is, any modifications on the bones happened as a result of the conditions experienced by the living animal. The baseline population available for use by a hunter, then, is a population of bones that is unmodified by extraneous agents and is complete with regard to the anatomy of the species hunted.

On the other hand, if an agent is scavenging, it is likely that the anatomy of the prey individual is modified away from its original form by virtue of consumption and transport of parts by the original predator, as well as by a series of other predator–scavengers feeding before the scavenger of interest selects parts for use. If we could specify the generalized composition of this predator–scavenger-modified population of anatomical parts remaining at kill–death sites, we would be in a position to know what was available for use by a late-arriving hominid scavenger.

In an earlier study (Binford 1981), I summarized information then available regarding the composition of faunal assemblages that survived the early stages of nonhuman predator–scavenger consumption. Since then a study by Richardson (1980a,b) supplies additional controlled observations on a number of facts of interest. I consider Richardson's study of major importance because he observed 89 carcasses over a considerable period of time, actually identifying the scavengers that fed on the carcasses. Richardson established blinds either adjacent to carcasses placed out as bait or near natural death sites so that he could observe the visitors to the site and the character of their consumption, as well as the consequences of feeding on the form of the remaining bones. Richardson observed changes in the survivorship of bones as a function of the duration of exposure to scavenger feeding. He also noted the sequence of modification or damage to different bony parts in terms of both the species differences among the observed agents and the intensity and diversity of feeders. Table 4.31 summarizes the frequencies of anatomical parts remaining on the carcass locations observed by Richardson after scavengers had ceased to feed. In addition, information is supplied regarding the percentage of the surviving bones that show damage in the forms of tooth scoring, pitting, and scooped-out areas of cancellous tissue; in other words, recognizable modifications typically produced by gnawing animals (for detailed descriptions, see Binford 1981: 35–86; Brain 1981:138–144; Maguire and Pemberton 1980). Figure 3.9 illustrates the relationship between the frequencies of parts remaining at the sites of scavenged carcasses, as documented by Richardson, and the population of parts remaining on sites obviously reported where primary predators had consumed parts at the sites of a successful kill (Binford 1981:230, Column 2). It should be clear that there is a strong, positive, curvilinear relationship between the two samples. The differences reflect differences in scale; that is, the scavenged carcasses are not as extensively exploited as are the carcasses representing primary kills by predators most of the time. On the other hand, the pattern of exploitation is roughly the same for primary predators and secondary scavengers. This generally means that, for a scavenger, a greater amount of usable food is available at sites of natural death if the scavenger can get there shortly after death, whereas parts of decreasing utility will be more generally obtainable from kill sites produced by active predators.

The next step in developing recognition criteria for a population of bones scavenged and subsequently introduced as consequences of hominid hunting efforts involves being able to (1) anticipate or demonstrate the bias in selection that the scavenger would be forced to follow, given the composition of the population available; and (2) demonstrate the state of the parts available for selection. What is the state of the bones identifiable by virtue of tooth marks, breakage, and so forth made before the hominid scavenger selected parts for use? I have in mind the fact that a carcass in an African setting is rather quickly attacked by flies, and will also be desiccated in the generally dry settings. This means that the carcass cannot be dismembered in the same way as a fresh carcass, because the mechanics are totally different for a stiff and dry carcass than for a fresh and supple one.

Returning to the original proposition, the baseline available for use by a *hunter* is a population of anatomical parts unmodified by extraneous agents, and complete as regards the anatomy of the species hunted. By way of contrast, we must emphasize that the baseline population available for

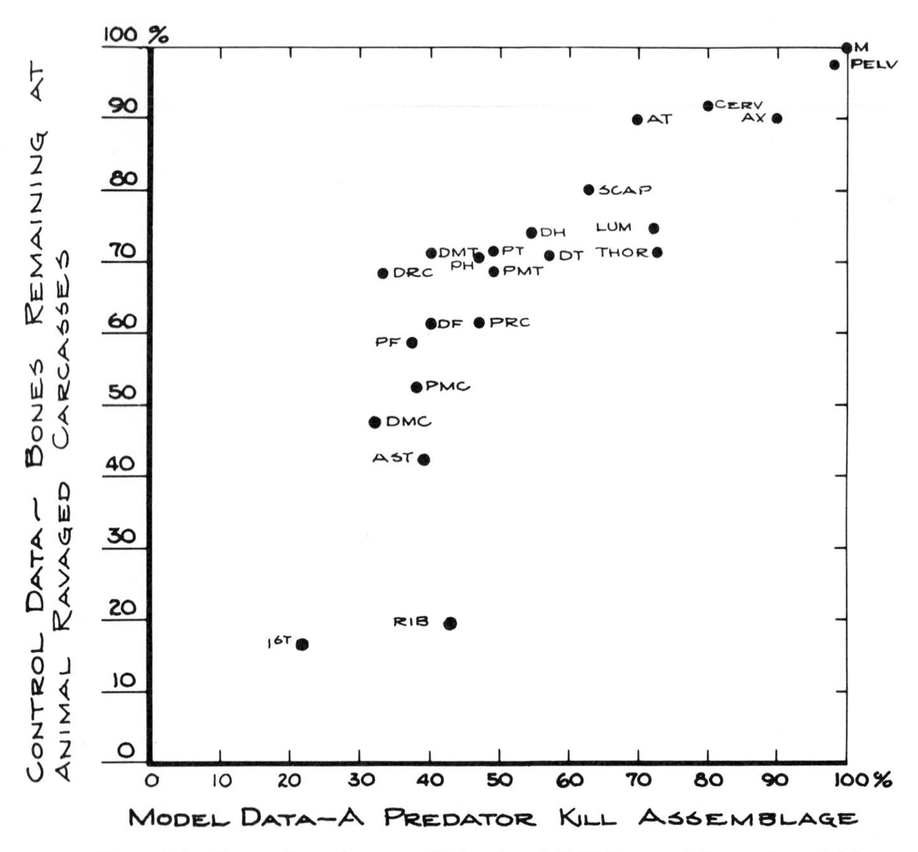

Figure 3.9 Comparisons between Richardson's (1980) control data and model data (Binford 1981: Table 5.08, Column 2) about the composition of scavenged carcass populations. AST, astragalus; AT, atlas; AX, axis; CERV, cervical vertebrae; DF, distal femur; DH, distal humerus; DMC, distal metacarpal; DMT, distal metatarsal; DRC, distal radiocubitus; DT, distal tibia; H, humerus; LUM, lumbar vertebrae; M, mandible; PELV, pelvis; PF, proximal femur; PH, proximal humerus; PMC, proximal metacarpal; PMT, proximal metatarsal; PRC, proximal radiocubitus; PT, proximal tibia; RIB, ribs; SCAP, scapula; 1st, first phalange.

use by a *scavenger* is a population of bones generally modified and ravaged by extraneous agents and surviving as an incomplete, modified, and biased inventory of anatomical parts. I have already presented data that provide a reliable description of the composition of this biased baseline population. Studies of wolves in Alaska and (mainly) lion kills in central Africa (Binford 1981), and now carcasses in a South African setting (Table 4.31) all converge to reinforce one another in justifying the generalization that predator–scavengers generally consume prey carcasses in a fairly regular sequence, so that the major differences in the anatomical inventory remaining from predation–scavenging at the kill locations are primarily a function of the intensity of the feeding and competition among consumers at the site. In short, we can specify fairly accurately the composition of the parent population from which a scavenger would have to draw his sample for either immediate consumption or transport elsewhere. (This model parent population is summarized in Binford 1981:Table 5.08, Columns 2 and 6.)

In addition to knowing the average form of the population of parts available to a scavenger, we also anticipate the parts within that population that a food seeker would tend to select. Stated another way, if we know the distribution of edible material on the anatomy of an ungulate, and we know something about the character of parts apt to survive initial feeding by nonhominid killers or other scavengers, we can anticipate the character of the population of parts most likely to be systematically recovered by a hominid scavenger for use as food (see, for instance, the model given in Binford 1978:188, Column 2).

Unfortunately, as is the case in many situations, the form of a faunal assemblage expected from scavenging is not totally unambiguous. There are known contexts in which modern humans may generate a population of anatomical parts that closely mimic the composition of a population of parts remaining on animal kill sites (see Binford 1978:Table 5.1; and 1981:233), as well as derived or transported samples from such a population (see Binford 1981:236). This means that, given our current levels of understanding, identification of the products of hunting versus scavenging cannot rest exclusively on facts of assemblage composition. There must be other properties of assemblages within this identifiable class of assemblage forms that, if known, would permit a resolution of ambiguities in identification.

I had the opportunity to observe a number of carcasses from both natural deaths and predator kills in the Nossob River area of South Africa and Botswana, as well as elsewhere. My observations sought out properties manifested on bone assemblages that would provide an observer with clues to the state of an anatomical part at the time it was transported or selected for use by a scavenger. As a result of these experiences, I reached two principal conclusions: (1) Knives are generally ineffectual tools for butchering or dismembering partially dry or semidesiccated carcasses. The skin is more of the consistency of wood than leather, and the bound condition of tendons (Figure 3.10) and ligaments "freeze" the joints, making butchering with cutting implements through the points of articulation very difficult. A combination of hacking and twisting is the technique that experimentally works best on stiff and dry carcasses. The use of a small hatchet or a machete for hacking at the dried tendon, and then twisting and pulling the

Figure 3.10 Ravaged wildebeest carcass in the Nossob River valley.

joints apart in conjunction with further hacking proved the most effective way of dismembering stiff and partially dry segments of the anatomy. The use of a knife for short, slicing cuts was almost totally ineffectual. This means that (2) the removal of anatomical segments from a carcass in a drying state, previously fed upon by other animals, is certain to result in a different pattern of cut, hack, and chop marks than will the butchering of a fresh carcass, even if a cleaver-type strategy is used.

A further set of observations were made in the field: the state of the carcass at the time of scavenging will condition where butchering or the cutting-off of parts will occur. This dismemberment focus relative to the gross anatomy of a dried carcass is different from dismemberment points selected when faced with a relatively complete, fresh carcass. For example, Figure 3.11 illustrates a mule-deer carcass that I observed within the upper Kootenee drainage of Montana. It had been extensively fed upon by coyotes, but was still essentially in a semifresh state in the sense that the bone marrow and some of the meat were still usable and had not yet began to exhibit evidence of putrefaction, although flies were present. What should be clear is that, at this stage, the parts that had not been consumed and ravaged were the lower legs and the head. The entire body cavity had been opened, and both muscle and soft tissue had been consumed by the coyotes. If I were a scavenger and decided to remove from this carcass parts with a marginal but still-consumable potential, I would focus my dismemberment strategies either at the metacarpal–radius joint or on the proximal tibia–distal femoral joint. Bones above these points in this case could be expected to exhibit tooth marks derived from carnivore activity. On other carcasses where consumption had not proceeded quite so far, disarticulation

Figure 3.11 Deer carcass fed upon by coyotes in Montana.

might be focused at the femoral–pelvic articulation and at the distal humerus–proximal radiocubitus junction. In any event, one could expect hacking and chopping to be manifest because the tough, attached skin and the stiff joints would render a knife essentially useless. Scavenging in the above case might also be characterized by an attempt to remove the head with its still usable tongue, soft fatty parts behind the eyes, and in other cancellous fossa of the skull. Field observations and experiment lead me to expect that if a tool-using scavenger dismembered parts remaining at sites of ravaged carcasses, dismemberment marks would be concentrated at articulations generally below ends and surfaces where evidence of carnivore gnawing might be located. In addition, those dismemberment marks effected on older carcasses would generally be chop and hack marks, indicative of coping with stiff and partially desiccated skin and tendon.

Finally, even when a carcass is not extensively dried but is already stiff, a number of problems are presented to a butcher. When butchering a supple carcass with tools, the joint may be manipulated to exert pressure on muscles and tendons, therefore rendering cutting a relatively easy task. But when a carcass is stiff, the joints are generally *bound*—the tissue has shrunk and locked the articulation into a fixed position, making manipulation of the joint impossible. This means that the orientation of cuts relative to the shape of bones will generally be in regular and determined places, rather than the more common situation in which the orientation of the cut shifts as the joint is flexed during dismemberment. There are many other mechanical consequences of a stiff versus a supple carcass, which I introduce as the specifics of anatomy are addressed. Suffice it to say that (1) patterned properties of placement and orientation to cut marks should aid in judging the state of the carcass at the time of dismemberment, and (2) dismemberment of parts during a scavenging episode can be expected to cope most often with a carcass that is stiff, with relatively inflexible joints.

Observations and field experiments lead me to such expectations. There was not, however, a body of controlled data from known scavenging contexts to use for describing exactly what form these expectations might take when seen in terms of statistical frequencies, as well as clustered associations of anatomical parts and tool–tooth-inflicted marks. On the other hand, I did have available samples from some Nunamuit Eskimo sites where the behavioral contexts were known and, in the light of the scavenging problems, could be studied to provide a control on what processing of fresh carcasses looked like when viewed in terms of dismemberment marks and marks derived from filleting meat. Table 3.3 summarizes observations made on a faunal assemblage collected at Anaktuvuk village in Alaska during 1971. This assemblage has been previously described (Binford 1978: 123–125, particularly Table 3.8, Columns 1 and 2) and represents the debris from processing essentially complete caribou carcasses for parts to be placed on drying racks. The parts were being dismembered in anticipation of the part to be dried. In addition, the parts of greatest food utility were being filleted so the meat could be dried in strip fashion. This means that filleting was concentrated on the parts of greatest utility as far as meat yields were concerned (see Binford [1978:15–45] for a discussion of these points).

Table 3.3 summarizes the cut marks observed on the bones recovered from this Eskimo dismemberment–filleting operation. The marks inflicted as a consequence of dismemberment acts are tabulated in Columns 3 and 4, whereas the marks inflicted during filleting operations (see Binford [1981] for a description of the two classes of marks) are presented on Columns 5 and 6.

Figure 3.12 illustrates the relations between the frequencies of these two classes relative to the gross anatomy of ungulates. Dismemberment marks are concentrated on the occipital condyles and the atlas vertebrae, deriving from the severing of the head by cutting from the dorsal surface just behind the skull into the articulation between the atlas and occipital condyles. Dismemberment marks are also present in high frequency on the pelves, deriving from cutting off the dislocated rear leg—something only really possible when a carcass is fresh. Dismemberment marks are also concentrated between the distal humerus and the proximal radiocubitus, and between the distal femur and proximal tibia. These are, in fact, the points of dismemberment observed during field butchering. Dismemberment was accomplished at these joints and then the meat filleted off the

Faunal Remains around Nunamiut Eskimo Ice Cellars[a]

Anatomical part	*MAU* (1)	*MNE* (2)	*No. marked (dismemberment)* (3)	% (4)	*No. marked (fillet)* (5)	% (6)	*No. marked (total)* (7)	% (8)
Antler	7	14	0	0	0	0	0	0
Skull	10	10	4	.40	3	.30	6	.66
Mandible	6	12	6	.50	1	.08	7	.58
Atlas	14	14	6	.43	0	0	6	.43
Axis	14	14	0	0	0	0	0	0
Cervical vertebrae	21	105	10	.10	24	.23	29	.28
Thoracic vertebrae	3	39	5	.13	28	.72	32	.82
Lumbar vertebrae	1	8	0	0	5	.63	5	.63
Innominate	1	2	1	.50	0	0	1	.50
Ribs	.5	12	3	.25	0	0	3	.25
Scapula	7	14	2	.14	7	.50	8	.57
Proximal humerus	23	46	4	.09	6	.13	8	.17
Distal humerus	23	46	15	.33	4	.09	17	.37
Proximal radiocubitus	13	26	4	.15	0	0	4	.15
Distal radiocubitus	10	20	1	.05	0	0	1	.05
Proximal metacarpal	9	18	2	.11	0	0	2	.11
Distal metacarpal	9	18	0	0	0	0	0	0
Proximal femur	24	48	3	.06	19	.40	21	.44
Distal femur	24	48	15	.31	18	.38	27	.56
Proximal tibia	22	44	16	.36	6	.14	18	.41
Distal tibia	12	24	3	.13	2	.08	4	.17
Tarsals	6	12	2	.17	0	0	2	.17
Astragalus	6	12	1	.08	0	0	1	.08
Calcaneus	6	12	0	0	0	0	0	0
Proximal metatarsal	6	12	2	.17	0	0	2	.17
Distal metatarsal	6	12	1	.08	0	0	1	.08
First phalange	9	72	0	0	0	0	0	0
Second phalange	9	72	0	0	0	0	0	0
Third phalange	9	72	0	0	0	0	0	0

[a] Processing area after spring hunting. Data from Binford (1978:124, Columns 1 and 2).

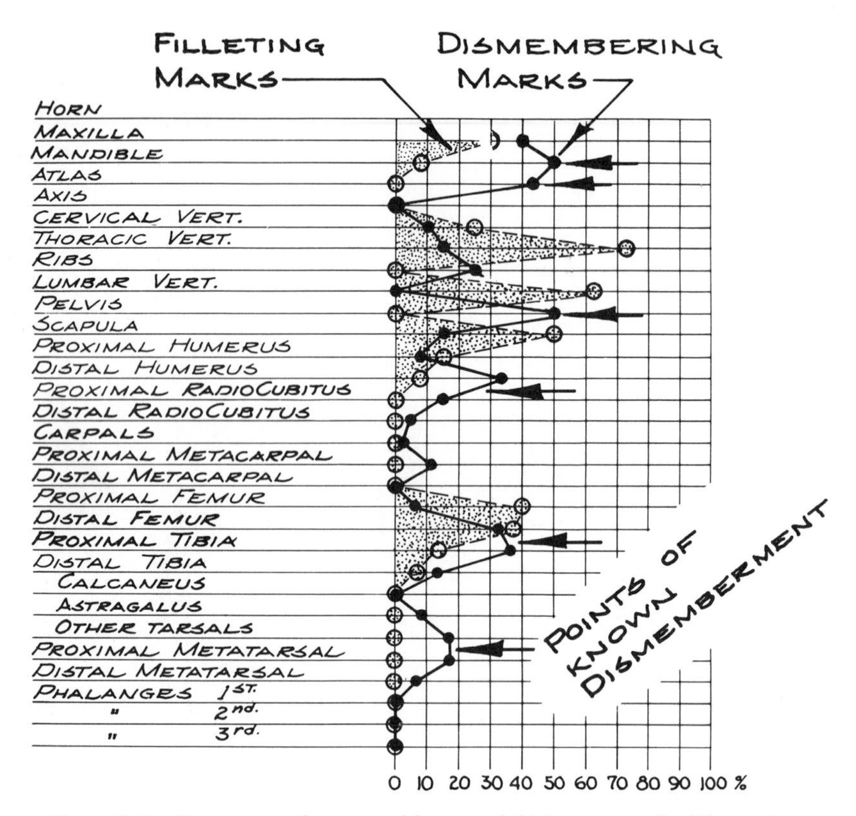

Figure 3.12 Percentage of recovered bones exhibiting cut marks (Nunamiut control data).

upper limbs for drying as strips, or left in a sheet attached to the scapula (see Binford [1978] for a description of these actions). Marks inflicted on the bones during filleting operations were therefore concentrated along the dorsal spines of the vertebrae, particularly the thoracic and lumbar vertebrae. These were inflicted at the time of the removal of the tenderloin. Others were on the scapula, where the meat of the shoulder was cut back from the bone, and on the femur, where the heaviest muscle mass is located on the caribou.

We may generalize that the position and frequency of marks inflicted during dismemberment are direct functions of the mechanics of dismemberment.

1. There is a direct relationship between the placement and frequency of tool-inflicted marks and the placement and frequency of dismemberment acts directed at particular anatomical locations.

2. There is a direct relationship between the placement and frequency of tool-inflicted marks and the placement and frequency of filleting acts directed at particular anatomical locations.

My work with the Nunamiut Eskimo and comparative studies of other faunal assemblages produced by fully sapient man show a further relationship between the focal areas customarily chosen for dismemberment and the food *utility* of parts being processed for transport and use. This principle applies equally to dismemberment and the parts selected for processing, such as the removal of meat in a filleting operation. Actions such as filleting will be concentrated on the anatomical parts yielding the greatest amount of desired product. This relationship was well illustrated in Figure 3.12, in which the concentrated frequencies were clearly understandable in terms of the utility of the parts being acted upon, such as the tenderloin, the shoulder, and the ham in the case of filleting. The frequencies of filleting marks may prove to be even more directly understandable than the butchering pattern, as shown by dismemberment marks. The greater security of the relationship between cutmark frequencies and processing investment derives from the fact that dismemberment may occur in different stages, and may reflect contingency-based decisions about transport somewhat independently of the simple decisions about what to use if faced with a complete carcass. This means that some or many inflicted dismemberment marks may refer to a series of transport or selective decisions made prior to processing. As a result, the relationship between dismemberment and filleting marks may appear partially incompatible, depending upon the dismemberment and disposal or abandonment decisions made prior to filleting. In any event, the processing traces in the form of tool-inflicted marks should inform us regarding the selection of anatomical parts for recovery and transport, and the processing of selected parts for use.

As can be seen, we have some guidelines as to the types of information that we might seek from a faunal assemblage in order to identify the consequences of hunted versus scavenged assemblages.

The knowledge basis from which I hope to develop inferential methods for identifying scavenging is roughly as follows:

1. We know something of the form of the population of parts remaining at carcasses that would be available for selection by a hominid scavenger.
2. We can assume that feeding is the scavenger's goal, and therefore the selection of parts from this remaining population would be in terms of food utility.
3. We know roughly the distribution of usable foods on the anatomy of ungulates, so we can anticipate the content of a population of

parts selected from a known baseline population in terms of criteria of food utility.

4. Experiences with carcasses have led to the realization that the mechanics of drying carcasses are different than the mechanics of fresh, supple carcasses. This being so, we can anticipate some differences in the kinds of tools regularly used and the dismemberment tactics employed, as they are directly conditioned by the state of the carcasses being dismembered.

In addition, I have some controlled data on cut marks from populations of animals hunted by man, and a general understanding of human hunting tactics. I know in some detail the difficulty of dismembering a dry carcass. On the other hand, I do not know the detailed character, or the specific placement, of marks resulting from repeated acts of dismembering dry carcasses. I also do not know the range of possible alternative ways of solving the stiff-carcasses problem. Put another way, I cannot go directly from my actual experiences to a realistic anticipation of the formal properties that might characterize an archaeological assemblage produced as a result of repeated acts of scavenging, in which the agents faced all the problems of selection and processing peculiarities outlined here.

In order to develop an unambiguous, expanded definition of scavenging that includes properties other than those of assemblage composition, I must work from both ends. As indicated, I have sought out actualistic experiences judged to be relevant to the properties potentially observable archaeologically. What is needed now is a good idea of the forms that these properties take when seen in archaeological materials. Since I have no directly controlled cases—as I did, for instance, with the dismemberment from hunted animals or from animal transported assemblages—the next best strategy is to seek out archaeologically recovered cases that meet all the formal properties of a partial definition for the recognition of a scavenged assemblage. This type of search considerably narrows the potentially relevant assemblages that could be studied for further insights into patterning possibly derived from scavenging. A selection from this group of provocative assemblages could then be made for detailed study. If, on detailed study, the formal properties of such attributes as breakage, animal gnawing, and cut-mark placement and forms could be shown to be consistent with the knowledge already available regarding the general types of patterns expected within a scavenged assemblage, we would have taken a first big step toward providing an expanded definition of scavenging for use in diagnosing other assemblages. Clearly, this strategy is only providing a watered-down link between the "bear and the footprint" (see Binford 1981:26–27), but it does provide detailed information about patterns in the archaeological record in terms of which more actualistic research could be planned. I am fairly convinced that we must work back and forth between dynamic and static experiences. This is because my own fieldwork has shown repeatedly that, when working in an actualistic context, I am frequently unaware of the character of static patterning; hence, I would be unable to seek out really useful information from actualistic experiences. Similarly, when I study the archaeological record for formal patterning, I almost always see patterns that are not understood. We must continually take knowledge of statics to dynamic experiences, and in turn bring a knowledge of statics to the search for formal patterns.

Middle Stone Age Anatomical-Part Frequencies from Klasies River Mouth Cave 1

A number of researchers have noted that although it is sometimes difficult to identify bones as to species, they can nevertheless be assigned to general categories such as *bovids,* and to different body-size classes. C. K. Brain (1969, 1974) was one of the first to suggest the systematic comparison among frequencies of bones from different body-size classes. Because Brain was interested in the diet of different African carnivores, he was well aware of differences among the various carnivores in the *prey image,* or the specific size and type of animal on which they preferentially preyed. Body size was thought to be an important variable, reflecting something of the food-procurement tactics of various predators.

Richard Klein basically accepted the size classes that Brain had previously suggested, with the added observation that the giant buffalo, *Pelorovis,* should perhaps be recognized as representing an independent body-size class because it was roughly double the size of the animals included in Brain's body-size class IV. The size classes into which the animals being studied have been grouped are:

Size Class I (Small Bovids) Class I includes the blue duiker, Cape grysbok, and oribi. These are all bovids weighing between 15 and 50 pounds. Their size range is indicated in Figure 3.13A.

Size Class II (Small–Medium Bovids) Class II accommodates vaalrhebok, mountain reedbuck, springbok, and bushbuck. The animals range between 70 and 110 pounds. The relative size is indicated by the scaled figure of a bushbuck (Figure 3.13B).

Size Class III (Medium–Large Bovids) Class III includes animals as small as approximately 150 pounds up to animals approaching 400 pounds in body weight. Typical animals include southern

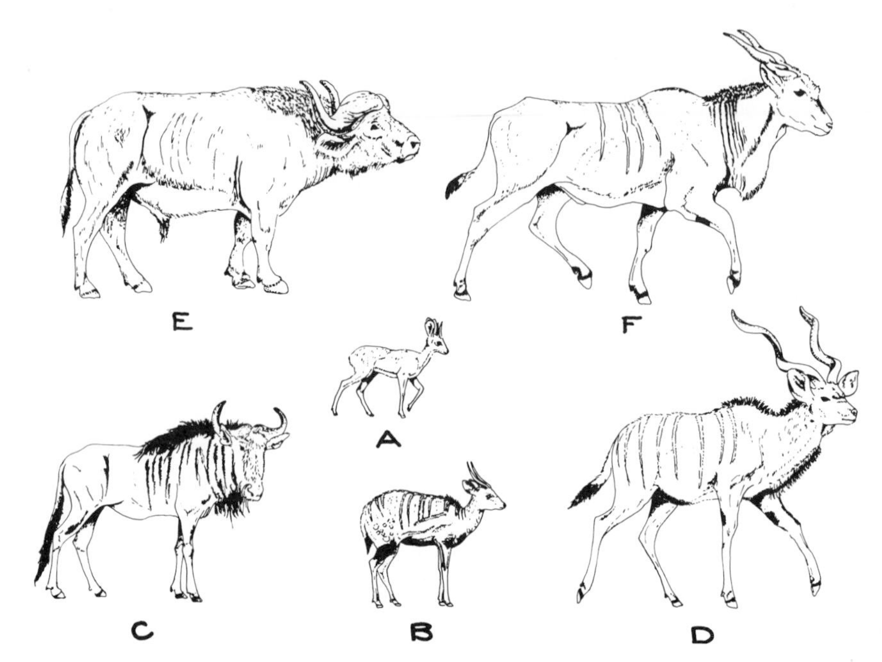

Figure 3.13 Scale comparison among species representing different body-size classes used in this study. (See text for identification of A–F.)

reedbuck, blue antelope, bastard hartebeest, hartebeest, wildebeest, and kudu. This class is represented by a blue wildebeest (Figure 3.13C) and the greater kudu (Figure 3.13D).

Size Class IV (Large Bovids) Animals in class IV are considered large and include the Cape buffalo and the eland (both 2000 pounds). These are shown on Figures 3.13E and 3.13F, respectively.

Size Class V (Very Large Bovids) In this analysis, body-size class V includes only the extinct giant buffalo, *Pelorovis,* which weighed approximately 4000 pounds.

Both the earlier study of the Klasies fauna by Klein (1976) and my restudy have employed these body-size categories as important analytic units.

Table 3.4 summarizes the information tabulated by Richard Klein (1976) using his observational conventions (MNIs), and Table 3.5 presents the same material tabulated by my observational conventions (MNEs and MAUs). Both tables are presented in terms of the five animal body-size classes used by Klein (1976) for describing the fauna in his initial studies.

TABLE 3.4

Tabulation of Klasies River Mouth Cave 1 Fauna by Klein

	Bovid class size									
	I Small[a]		*II Small–medium*[b]		*III Medium–large*[c]		*IV Large*[d]		*V Very large*[e]	
Anatomical part	*MNI (1)*	*% (2)*	*MNI (3)*	*% (4)*	*MNI (5)*	*% (6)*	*MNI (7)*	*% (8)*	*MNI (9)*	*% (10)*
Horn										
Occipital condyle										
Maxillary arc										
Maxillary teeth	27	.50	16	.34	52	.73	93	.72	21	.45
Mandibular teeth	46	.85	35	.77	71	1.00	129	1.00	47	1.00
Atlas	10	.18	5	.11	15	.21	8	.06	6	.13
Axis	12	.22	11	.24	7	.10	13	.10	2	.04
Cervical vertebrae	8	.15	9	.20	11	.15	13	.10	6	.13
Thoracic vertebrae	9	.17	6	.13	9	.13	7	.05	4	.09
Lumbar vertebrae	9	.17	9	.20	8	.11	10	.08	2	.04
Innominate	30	.56	21	.67	22	.31	17	.13	7	.15
Scapula	54	1.00	46	1.00	41	.58	18	.14	0	0
Proximal humerus	4	.07	6	.13	4	.06	6	.05	1	.02
Distal humerus	24	.44	18	.39	18	.25	19	.15	9	.19
Proximal radiocubitus	10	.18	8	.17	11	.15	33	.26	9	.19
Distal radiocubitus	6	.11	6	.13	10	.14	15	.12	4	.09
Carpals	0	0	2	.04	11	.15	49	.38	15	.32
Proximal metacarpal	4	.07	4	.09	21	.30	49	.38	13	.28
Distal metacarpal	3	.06	6	.13	10	.14	26	.20	6	.13
Proximal femur	14	.26	9	.20	7	.10	15	.12	3	.06
Distal femur	13	.24	12	.26	9	.13	12	.09	3	.06
Proximal tibia	12	.22	7	.15	8	.11	3	.02	1	.02
Distal tibia	13	.24	14	.31	27	.30	32	.25	5	.11
Tarsals	5	.09	2	.04	12	.17	29	.23	14	.30
Astragalus	9	.17	12	.26	32	.45	54	.42	10	.21
Calcaneus	20	.37	16	.35	22	.31	31	.24	9	.19
Proximal metatarsal	6	.11	8	.17	17	.24	41	.32	11	.23
Distal metatarsal	8	.15	8	.17	8	.11	10	.08	6	.13
First phalange	5	.09	8	.17	11	.15	27	.21	11	.23
Second phalange	5	.09	8	.17	11	.15	27	.21	11	.23
Third phalange	5	.09	8	.17	11	.15	27	.21	11	.23

[a] Steenbok, grysbok, oribi (Klein 1976:Table 9).
[b] Springbok, mountain redbok, bushbok (Klein 1976:Table 10).
[c] Blue antelope, kudu, hartebeest, bastard hartebeest, wildebeest (Klein 1976:Table 11).
[d] Cape buffalo, eland (Klein 1976:Table 12).
[e] Giant buffalo (Klein 1976:Table 13).

TABLE 3.5

Tabulation of Klasies River Mouth Cave 1 Fauna by Binford

	Bovid class size														
	I Small			II Small–medium			III Medium–large			IV Large			V Very large		
Anatomical part	*MNE* (1)	*MAU* (2)	% (3)	*MNE* (4)	*MAU* (5)	% (6)	*MNE* (7)	*MAU* (8)	% (9)	*MNE* (10)	*MAU* (11)	% (12)	*MNE* (13)	*MAU* (14)	% (15)
Horn	22	11.0	.21	9	4.5	.18	18	9.0	.23	23	11.5	.19	0	0	0
Occipital condyle	4	2.0	.04	9	4.5	.18	8	4.0	.10	26	13.0	.21	2	1.0	.04
Maxillary arc	22	11.0	.21	14	7.0	.28	9	4.5	.11	7	3.5	.06	6	3.0	.13
Maxillary teeth[a]	136	13.0	.25	92	9.0	.36	237	23.0	.58	374	37.0	.60	858	8.5	.38
Mandibular teeth[b]	324	29.0	.56	229	21.0	.84	436	40.0	1.00	583	62.0	1.00	212	22.5	1.00
Mandible	30	15.0	.29	27	13.5	.54	25	12.5	.31	32	16.0	.26	21	10.5	.47
Atlas	7	7.0	.14	5	5.0	.20	4	4.0	.10	7	7.0	.11	4	4.0	.18
Axis	12	12.0	.23	7	7.0	.28	6	6.0	.15	13	13.0	.21	4	4.0	.18
Cervical vertebrae	20	4.0	.08	30	6.0	.24	47	9.4	.24	36	7.2	.12	10	2.0	.09
Thoracic vertebrae	64	4.9	.10	69	5.3	.21	88	6.8	.17	55	4.2	.07	9	0.7	.03
Lumbar vertebrae	42	6.0	.12	42	6.0	.24	52	7.4	.19	55	7.8	.13	12	1.8	.08
Innominate	21	10.5	.20	23	11.5	.46	19	9.5	.24	17	8.5	.14	6	3.0	.13
Scapula	103	51.5	1.00	50	25.0	1.00	61	30.5	.76	35	17.5	.28	7	3.5	.16

Proximal humerus	6	3.0	.06	4	2.0	.08	4	2.0	.05	7	3.5	.06	2	1.0	.04
Distal humerus	37	18.5	.36	40	20.0	.80	26	13.0	.33	27	13.5	.22	9	4.5	.20
Proximal radiocubitus	12	6.0	.12	14	7.0	.28	12	6.0	.15	24	12.0	.19	11	5.5	.24
Distal radiocubitus	2	1.0	.02	8	4.0	.16	15	7.5	.19	20	10.0	.16	2	1.0	.04
Carpals	0	0	0	0	0	0	58	5.8	.15	159	15.9	.26	21	2.1	.09
Proximal metacarpal	5	2.5	.05	11	5.5	.22	20	10.0	.25	51	25.4	.41	9	4.5	.20
Distal metacarpal	4	2.0	.04	10	5.0	.20	17	8.5	.21	50	25.0	.40	8	4.0	.18
Proximal femur	18	9.0	.17	11	5.5	.22	15	7.5	.19	25	12.5	.20	5	2.5	.11
Distal femur	17	8.5	.17	14	7.0	.28	7	3.5	.09	14	7.0	.11	1	0.5	.02
Proximal tibia	13	6.5	.13	10	5.0	.20	19	9.5	.24	7	3.5	.06	2	1.0	.04
Distal tibia	22	11.0	.21	21	10.5	.42	45	22.5	.56	41	20.5	.33	6	3.0	.13
Astragalus	17	8.5	.17	21	10.5	.42	42	21.0	.53	68	34.0	.55	21	10.5	.47
Calcaneus	25	12.5	.24	20	10.0	.40	34	17.0	.43	31	15.5	.25	21	10.5	.47
Proximal metatarsal	7	3.5	.07	21	10.5	.42	15	7.5	.19	39	19.5	.31	14	7.0	.31
Distal metatarsal	12	6.0	.12	18	9.0	.36	14	7.0	.18	26	13.0	.21	16	8.0	.36
First phalange	8	1.0	.02	30	3.75	.15	60	7.5	.19	126	15.75	.25	314	3.9	.17
Second phalange	1	0.13	.002	3	0.38	.02	24	3.0	.08	112	14.0	.23	28	3.8	.16
Third phalange	0	0	0	4	0.3	.02	38	4.75	.12	120	15.1	.24	296	3.7	.16

[a] A convention was used here of 10.1 teeth equals one animal unit. This is obviously lower than the 12 cheek teeth that an adult bovid has. Experimental data show that for an "adult" biased population, this is the empirical average of maxillary teeth per individual.

[b] A convention of 9.4 was used for the mandible, since experimentally this was the best observed-value. The difference is probably related to variations in the likelihood of recovery for maxillary premolars as opposed to mandibular teeth.

Even the causal reader will notice that there is a considerable lack of identity between the frequencies (MNIs) tabulated by Klein and the frequencies (MAUs) tabulated for the same bones by me. For instance, Klein tabulates 129 MNIs respresented in the large-bovid body-size class IV (Table 3.4 Column 7), whereas I report only 62 animal units for the same material. This difference is perhaps the most striking between Klein's data and mine. In order to understand such discrepancies, we must return to a comparison of our methods.

Consistently, the greatest differences between my tabulations and Klein's are in terms of units based on teeth. Klein sought to estimate the minimal number of animals that would have died to account for the teeth variable, in terms of age, rights and lefts, and specific identification. For instance, if one immature right and four adult left third-premolars were observed, Klein would enter an MNI estimate of five animals. My approach would lead to an estimate of 2.5 maxilla or skulls, the element I believe to have been most likely selected by past humans for transport and introduction to a site. If people of the past had in fact introduced split skulls for some reason from four adults, the differences between right and left are not known to make any difference in terms of consumer goods represented. They would as consumer goods be equivalent to two complete skulls, which is the type of information I seek: How did people make transport, processing, and consumption decisions with regard to various segments of an animal's anatomy?

There is another reason for differences between Klein's data and mine. He does not use fractional MNI estimates, whereas I use fractional MAU estimates. If I have only one proximal humerus, it is listed as 0.5 animal units, whereas Klein would see it as representing at least one dead animal. This plays a considerable role in generating differences between our summary counts. Because Klein never considers that a proximal femur from one level and one from another level, taken together, represent only one animal unit, he sees the two femur heads as representing at least two animals killed at different times. Thus, when adding the MNIs from two separate levels, two proximal femur heads would sum to two individuals, whereas my method would produce a MAU of only one individual unit. This difference contributes strongly to some of the differences noted among our respective data sets.

Our methods, when understood, are in fact differing ways of looking at the same reality, that is, fragmentary and partial bones from fragmentary and partial anatomies expressed in terms of a complete and conveniently segmented anatomy. To illustrate this point, I have plotted my percentage indices against the indices from Klein's MNIs for two body-size categories,

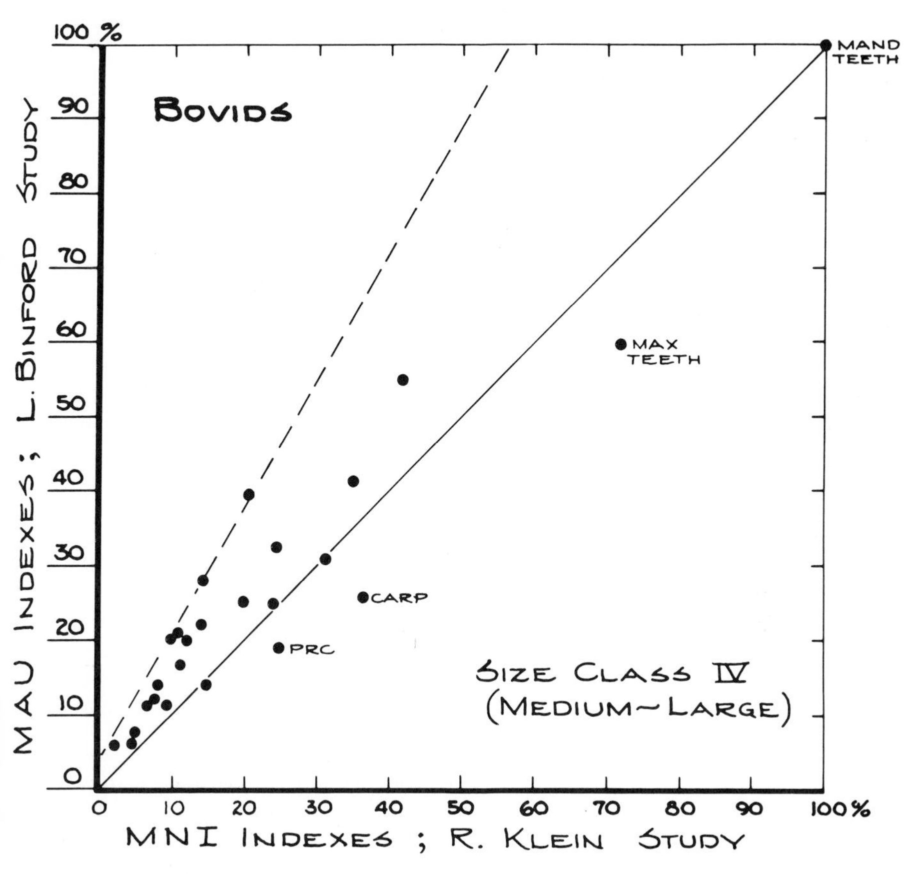

Figure 3.14 Comparison of Binford's and Klein's tabulations of anatomical parts for large-size bovids. CARP, carpals; MAND, mandibular; MAX, maxillary; PRC, proximal radiocubitus.

large (Figure 3.14) and small (Figure 3.15). From this, it is clear that the tabulations are strongly correlated but estimates based on teeth contrast most, while innominate parts among small animals seem also to be considerably different. I suspect that Klein did not go through the step of estimating MNEs as opposed to simply counting the fragments of recognizable innominates in arriving at his estimates. In any event, there are differences between the tabulations that are to be expected, given the different conventions used in observations and reporting. These differences result in different tabular figures. However, the overall pattern of survivorship at Klasies seems to be reflected well by both approaches.

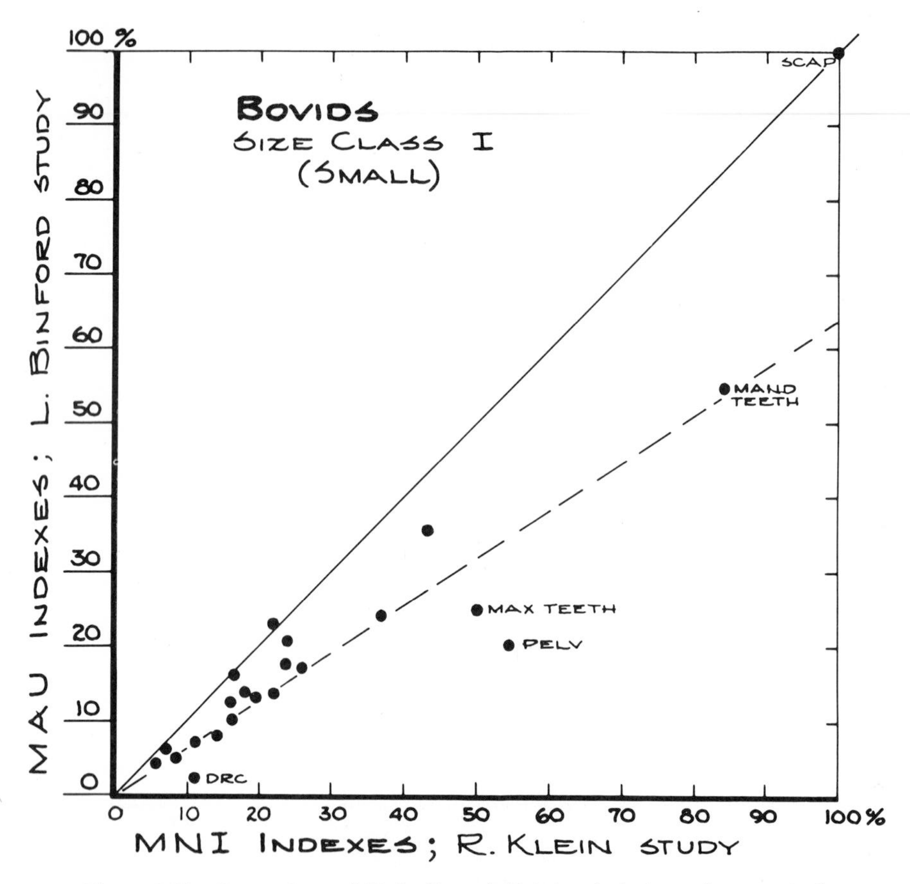

Figure 3.15 Comparison of Binford's and Klein's tabulations of anatomical parts for small bovids. DRC, distal radiocubitus; MAND, mandibular; MAX, maxillary; PELV, pelvis; SCAP, scapula.

Interpretation of Patterning

Klein emphasized differences in the anatomical-part frequencies characteristic of small- versus large-body-size animals. My study sustains the difference and even amplifies the pattern. Figure 3.16 illustrates the MAUs tabulated for small versus large bovids from Klasies River Mouth Cave 1.

A substantial group of body parts from both large and small animals are moderately to poorly represented in both populations. These appear to be linearly and positively correlated. These parts are the atlas; axis; cervical, thoracic, and lumbar vertebrae; as well as the pelvis—in other words, the

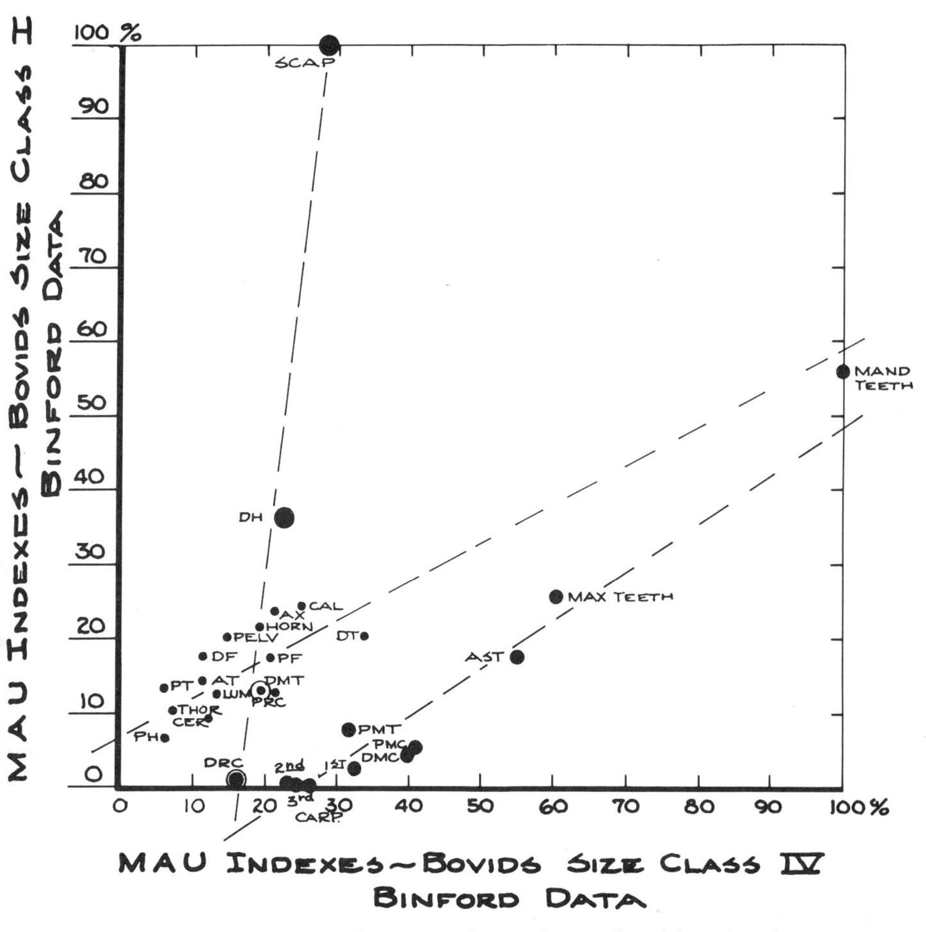

Figure 3.16 Comparison of anatomical parts for small and large bovids. AST, astragalus, AT, atlas, AX, axis; CAL, calcaneus; CARP, carpals; CER, cervical vertebrae; DF, distal femur; DH, distal humerus; DMC, distal metacarpal; DMT, distal metatarsal; DRC, distal radiocubitus; DT, distal tibia; LUM, lumbar vertebrae; MAND, mandibular; MAX, maxillary; PELV, pelvis; PF, proximal femur; PH, proximal humerus; PMC, proximal metacarpal; PMT, proximal metatarsal; PRC, proximal radiocubitus; PT, proximal tibia; SCAP, scapula; 1st, 2nd, 3rd, phalanges.

entire axial skeleton except the head. As has been shown repeatedly, these are the parts that commonly remain after feeding at predator kills unless there is intense competition among feeders (Binford 1981; Hill 1975, 1979; Mech 1966; Richardson 1980b; Shipman and Phillips-Conroy 1977). It appears that this complete axial skeleton unit was frequently not returned to the site, but when this did occur, the pattern of parts returned was roughly the same for small as well as larger animals. The axial skeleton, proximal and distal femur, proximal and distal tibia, and the calcaneus were represented. These are parts that would generally remain attached by skin to a lightly ravaged carcass. Apparently, whatever it is that conditions the relative frequencies of these parts at the Klasies Cave 1 site may well be common to both the large and small animals. On the other hand, parts frequent in the small-bovid population and only moderately represented in the large-bovid population are the scapula and the distal humerus, the meat-yielding parts of the upper-front leg. It must be kept in mind that the scapula and upper-front limb constitute one of the easiest anatomical segments for carnivores to disarticulate (see Figure 3.10). This means that it is also a part commonly removed from the kill–death site by wary scavengers (see Shipman and Phillips-Conroy 1977; Table 1) and is frequently the part commonly cached by predator–scavengers. Some kind of selective use of body parts is clearly indicated by the contrasts between the two graphs. The challenge is to isolate the source of the bias.

Parts very common in the large-bovid population and poorly or moderately represented among the small bovids are the mandible and maxilla (the head parts), as well as tarsals, astragalus, metatarsal, phalanges, and distal radiocubitus carpals and metacarpals; in other words, the lower-front limb-bones and the lower-rear leg. The former is most often butchered through the shaft of the radiocubitus, whereas the latter is most often removed by cutting between the astragalus and the calcaneus. In short, the bias present among the bones from the large animals is the very pattern that attracted me to the study of the material—a pattern of high head-and-lower-limb-bone frequencies, the pattern that all actualistic evidence suggested should be characteristic of a scavenged assemblage.

Given the aim of interpreting the differential part frequencies exhibited between the animals of different body size, the first problem that must be addressed is the possible role of differential survival of anatomical parts both within a body-size class and among the classes compared. I have shown that a population of bones subjected to destructive agents may be so modified with respect to surviving part frequencies that the original composition of the population may be rendered unrecognizable, whereas its derived form may be only referable to the differential durability of the different bones themselves.

In terms of the tests previously described (Binford 1981:217–222), the population of bones recovered from Klasies River Mouth has suffered from selective deletions as a function of the relative durability of bones. Figure 3.17 illustrates the survival ratios between proximal and distal ends of the humerus and tibia. For both bones, all the faunal populations show that they have been modified away from the proportions present in a live animal

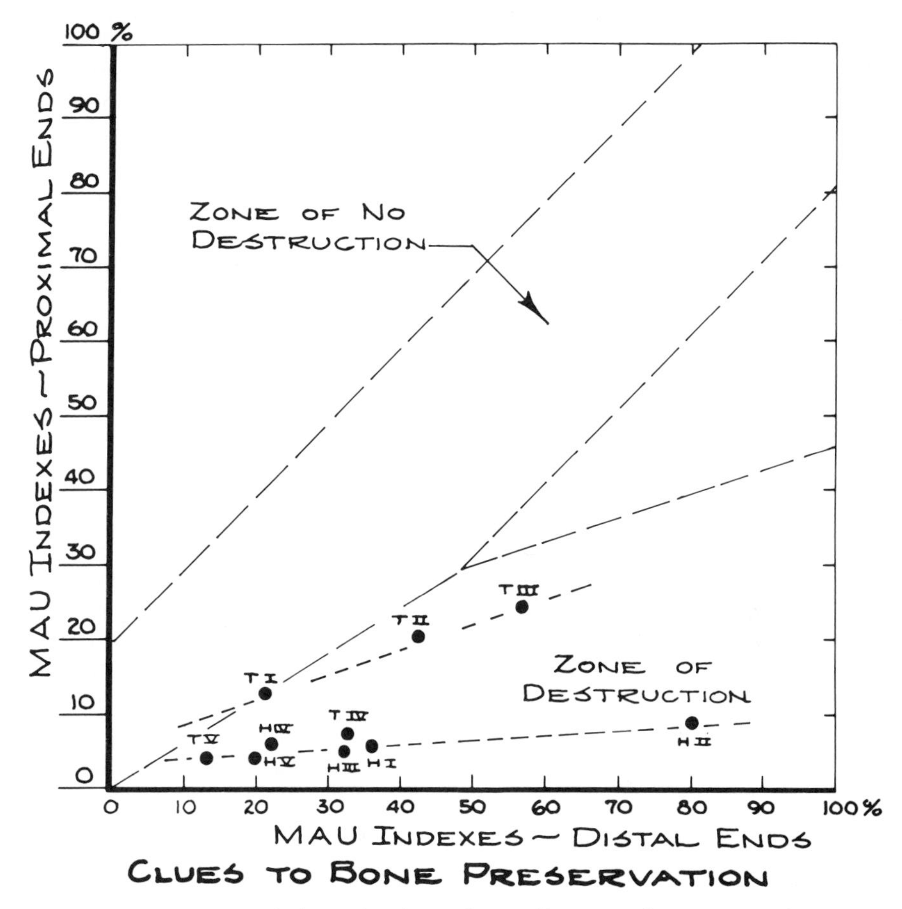

Figure 3.17 Test of Klasies data for evidence of bone-part destruction. H, humerus; T, tibia. Roman numerals indicate body-size classes.

as a simple function of the relative durability of the bone parts themselves. Soft bones are poorly represented whereas hard, dense bones are well represented.

Experimental study of the bone density for sheep and caribou from individuals of differing ages has provided a body of data that can be used to estimate the survival potential of different bones in the anatomy. I have previously described this work (Binford 1978:210–211; Binford and Bertram 1977) and will use these figures to correct the surviving population for the effects of ravaging agents. By dividing the observed MAU by the experimentally established survival percentage (Binford 1981:218), we obtain an estimate of the MAU originally present prior to the action by ravaging agents. This reconstructured assemblage removes the ambiguity created by the interaction between the form of the original population and attritional agents acting on that population. Table 3.6 presents the reconstructed values for the small bovids (Table 3.5, Column 2) and the large bovids (Table 3.5, Column 11). The first condition that is clarified by the reconstruction is that the small-bovid population exhibits a segmental structure that makes good anatomical sense. This segmental structure clearly reflects the anatomical segments in terms of which the gross anatomy of the small bovids was treated, and in turn identifies the anatomical units in terms of which the bones were actually introduced to the site.

Inspection of Table 3.6 shows that there are basically seven anatomical segments (see Binford 1981:91–95) that behave as units quantitatively distinct one from another. First is the head-and-upper-neck unit, represented primarily by the mandible and the atlas and axis vertebrae. The low frequency of the cranium within this unit, represented by maxillary teeth, may indicate at least a partial destruction of the cranium in the field before the head and upper neck were transported to the site. The next unit is the thorax and spine. This is the anatomical unit least frequently introduced to the site. It should be recalled that this is one of the anatomical segments most commonly remaining at abandoned animals kills after predator–scavengers have finished feeding (see Binford 1981:230). By way of contrast, these parts were regularly transported for storage by the Nanamiut Eskimo (Binford 1978:112). In short, high frequencies of these parts commonly remain where gross consumption takes place, or where one is forced to make hard transport decisions carrying only the parts of highest utility (see, for instance, my discussions of piece butchering, Binford 1978:60–64). Under such contingencies, these parts would be abandoned at procurement locations. It is clear that they were rarely introduced to Klasies River Mouth.

Given what we know about animal behavior and about human hunting that is logistically organized—that is, where food is transported to consumers located at places other than the locations of procurement—it would appear that primary consumption of the small antelope did not take place at the Klasies River Mouth site. That is, small-bovid parts were not introduced as complete animals but instead as segments removed from animals killed, and were either piece butchered or partially consumed at some place other than the site at Klasies River Mouth.

When we turn to the segment designated the upper-front leg, a pattern of tactical significance is implied. Unlike the spine and thorax, where each anatomical subunit part is represented in roughly equal frequencies, the upper-front leg is represented by scalar frequencies, with the scapula absolutely most common and each bone lower down the leg represented by

TABLE 3.6

Reconstructed Frequencies for Small and Large Bovids[a]

Anatomical segment		Small bovids MAU (1)	Small bovids % (2)	Large bodids MAU (3)	Large bodids % (4)
	Horn	0	0	0	0
Head and upper neck	Maxilla	13.0	.21	37.0	.40
	Mandible	29.0	.46	62.0	.67
	Atlas	22.1	.35	12.0	.13
	Axis	27.1	.43	29.0	.31
Thorax and spine	Cervical vertebrae	11.4	.18	20.0	.22
	Thoracic vertebrae	12.3	.20	10.5	.11
	Lumbar vertebrae	11.8	.19	15.29	.17
	Pelvis	13.1	.21	10.63	.11
Upper-front leg	Scapula	62.8	1.00	21.34	.23
	Proximal humerus	30.0	.48	35.0	.38
	Distal humerus	22.84	.36	16.67	.18
Lower-front leg	Proximal radiocubitus	7.5	.12	15.0	.16
	Distal radiocubitus	1.67	.03	20.0	.22
	Proximal metacarpal	8.3	.13	84.7	.91
	Distal metacarpal	4.3	.07	53.2	.57
Upper-rear leg	Proximal femur	15.5	.25	21.5	.23
	Distal femur	18.48	.29	15.2	.16
	Proximal tibia	15.85	.25	8.5	.09
	Distal tibia	14.85	.24	27.7	.30
	Calceneus	18.0	.29	22.8	.25
	Astragalus	14.91	.24	59.7	.64
Metapodials	Proximal metatarsal	9.21	.15	51.3	.55
	Distal metatarsal	20.0	.32	43.3	.47
Phalanges	First phalange	5.8	.09	92.6	1.00
	Second phalange	1.1	.01	92.6	1.00
	Third phalange	0	.00	92.6	1.00

[a] Values obtained by multiplying Table 3.5, Columns 2 and 11 by values given in Binford (1981:Table 5.04, Column 4).

decreasing frequencies until we approach the proximal radiocubitus. The latter is represented by only 12% of the number of scapulae in the population. This type of pattern is derived from a selection process in which the decision to butcher off parts of lower food-value varies situationally with the decision to carry back the shoulder or the meat-yielding upper-front-leg unit. This decision was even made with respect to the bones of the upper leg. The lower-front-leg segment, from the radiocubitus down, is marginally represented by about 12 to 23% relative to the count of the scapulae. The low frequencies of the distal radiocubitus and distal metacarpal is probably a function of variability in survival probabilities associated with small animals not accurately anticipated by the survival percentages used in this reconstruction.

Among the Nunamiut Eskimo, the lower-front leg was frequently removed as a unit and either discarded or consumed in hunting camps and stations (Binford 1978:62–64). Clearly an analogous removal is indicated by these data. It should be pointed out that this patterned frequency break between the distal humerus and proximal radiocubitus in an equivocal assemblage is probably sufficient to suggest strongly that human agents were responsible for dismemberment, since animals tend to destroy the proximal humerus and chew down the humeral shaft, leaving the distal humerus attached to the proximal radiocubitus.

The frequency-patterning characteristics of the axial unit and the upper-front and -rear legs seem anomalous in that the occupants of the cave were hunting these small bovids and returning the kills (which are easily transportable as whole animals) back to the home base for consumption. The pattern of parts returned is most analogous to a *piece-butchering* strategy in which an animal is killed and only a few parts are returned to the residential base camp. Among modern hunters with whom I am familiar—those having a strong division of labor and typically an ethic of sharing hunted foods—piece butchering is only practiced when transport of the entire kill is delayed, so that the introduction of the complete kill at a later time is anticipated. This means that in terms of anatomical-part frequencies, this behavior is really only clearly visible among kill-site assemblages in which the anticipated future transport did not take place. Since this is a rare situation, the piece-butchered contribution is rarely visible in base-camp faunal populations, because most of the time the usable parts of the kill would have been introduced eventually.

For the results of piece butchering to be as obvious and visible as is seemingly the case at Klasies River, it would have to be a regular action. This would imply that the hunters regularly had a transport problem, which seems highly unlikely, because the bovids being discussed are so small that they can be carried easily by a single individual (see Figure 4.21). The only other context in which I can imagine a regular and patterned biased introduction of meat-yielding parts from the upper-front leg is if the consumption of the choice parts of the carcass had taken place at the kill or point of procurement. Under these conditions, transport of parts from the kill would be biased in favor of parts of secondary importance, which the shoulder certainly is, relative to the rear leg.

Before going on with the development of an accommodative behavioral model, we must complete the description of the patterning. In this light the patterning manifested among the parts of the rear leg is really quite interesting. This is the part of an ungulate with the greatest amount of usable or consumable meat (see Binford 1978:17–19). The pattern is very clear that all bones of the upper-rear leg were introduced in equal frequencies—indicating that when the upper-rear legs were introduced, they were complete upper legs disarticulated between the astragalus and the navicular cuboid, or between the navicular cuboid and the proximal matatarsal. These complete upper-leg units were carried back to the site, as opposed to the situation with the upper-front leg, which was segmented to carry only the scapula back and to remove the humerus in the field about 50% of the time. Whenever the upper-rear leg was introduced, it was carried back complete. The upper-rear leg was returned only about 25% as commonly as was the upper-front leg, in spite of the fact that the rear legs have more edible meat. This provocative fact appears even more anomalous when it is realized that there is a greater proportion of the total rear-leg meat on the femur, while on the front leg there is a more equable distribution of meat shared between the humerus and the scapula. Put quite simply, if one were making a decision as to which upper leg to disarticulate and abandon, while maximizing the amount of usable meat remaining on a single bone, the rear leg would certainly be the one chosen and the femur would be the bone transported. But the pattern observed among the small-bovid fauna at Klasies River Mouth is just the opposite. Upper-rear legs were rarely returned to the site, and when they were, they were introduced as complete upper-leg segments. On the other hand, the upper-front leg was the part most commonly returned to the site and it was most commonly introduced after being culled to the most productive single bone—the scapula. The pattern of the rare introduction of the metapodials and phalanges is totally consistent with the practices of modern hunters, where there is a biased abandonment or consumption of these low-yield parts in the field near the kill or hunting location. (See Binford [1978:62–64] for a description of the biased abandonment of lower-leg parts at kill sites.)

Given what is known about animal anatomy and the transport of anatomical parts by both humans and animals, the patterns manifested by the small-bovid bones seems quite straightforward:

1. The rather robust segmental patterning into units disjointed between articular surfaces is totally consistent with tool-using hominids as the likely dismembering agents.
2. The high relative-frequency of mandibular and upper-neck parts along with an absolute bias in favor of the scapula betrays a biased introduction of these anatomical segments to the site.

The latter parts have only moderate utility. Piece butchering such as this is normally the result of a transportation problem, meaning that the food available is present in greater quantity than can be carried by the field party. Clearly with such small bovids this cannot be the situation with the Klasies animals. In addition, under conditions of transport-related piece butchering and the biased selection of parts, those generally chosen for transport are the ones of greatest utility. At Klasies we see a consistent bias in favor of selective transportation of moderate-to-marginal parts (e.g., the neck). This could only be expected if the parts of highest utility were not available due to prior consumption at the point of kill by hominid or other animal agents.

The regular segmentation of the carcass strongly suggests that other animals were not involved in dismemberment, leading me to suspect that most of the small animals represented at Klasies River Mouth were eaten by hominids at the kill or death locations prior to the return of selected parts to the Klasies River Mouth site. The differences in patterning manifested by the prime anatomical parts (rear leg and lower back) is consistent with a model (see Figure 3.18) of the formation of the small-bovid population in which about 25% of the animals represented were introduced to the site "whole" (frequently with the lower-front leg and the metatarsals plus phalanges removed), while 75% of the small-bovid animal units were introduced to the site after the consumption of prime parts elsewhere, presum-

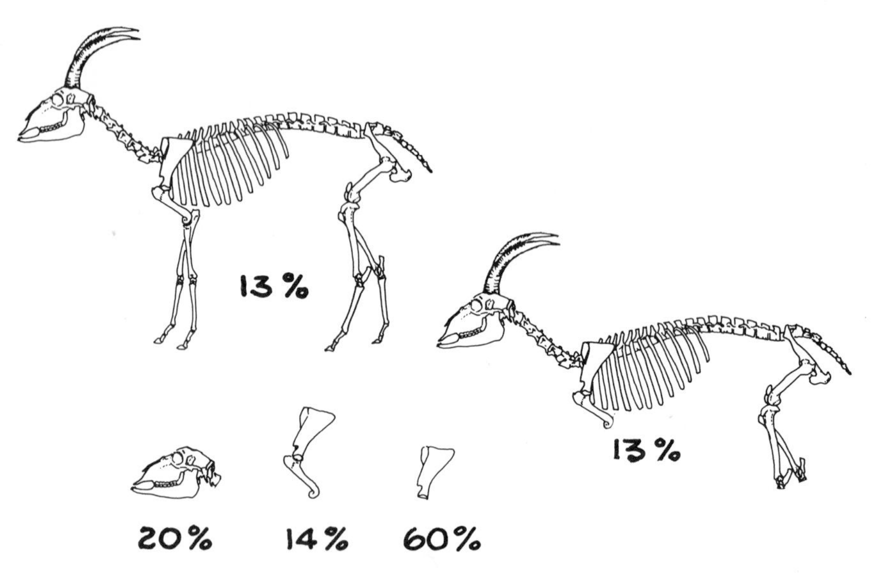

Figure 3.18 Relative frequencies of segments of small bovids most often introduced to the site.

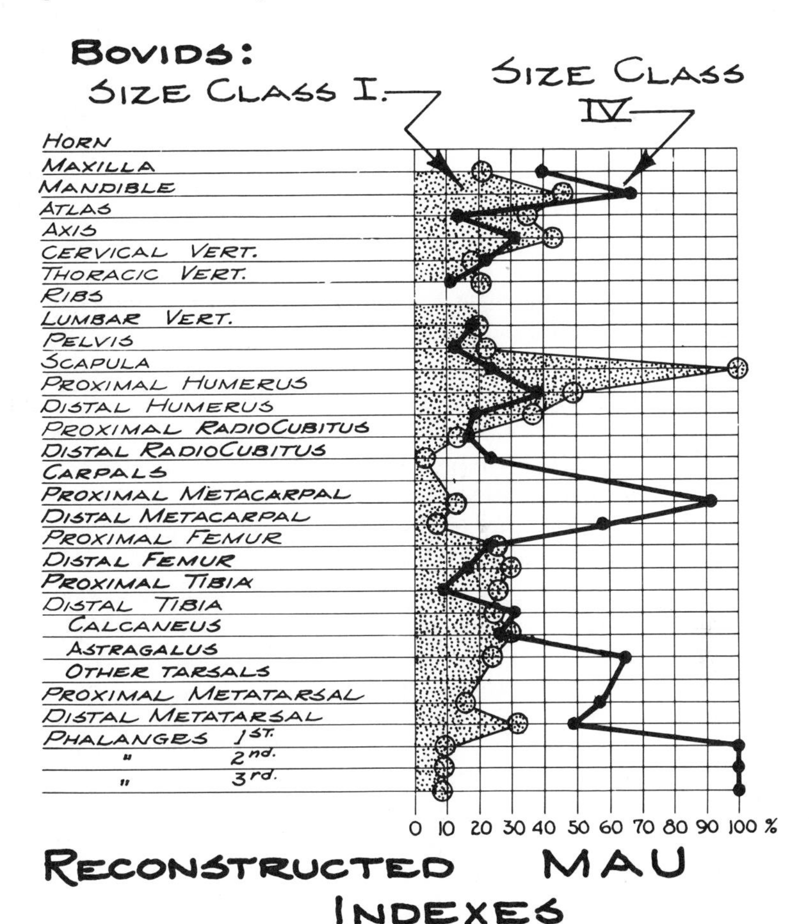

Figure 3.19 Comparison between anatomical part frequencies from small and large bovids (reconstructed values).

ably at the kill–find location. In any event, primary consumption elsewhere seems indicated.

Turning now to the body-part frequencies from the larger bovids (Table 3.6), we note an almost total contrast in all properties of the anatomy represented (Figure 3.19). First, the most common parts are the phalanges and metapodials—parts that we have seen to be commonly abandoned by modern hunters as of marginal utility and hence worthy of little investment as far as transport is concerned. At Klasies River Mouth these parts of low general utility are the parts most commonly introduced to the site from the larger animals.

Turning to the bones of the upper limbs, the meat-yielding anatomical segments, we note a scalar pattern for the upper-front leg with a positive bias in favor of the scapula and a scalar pattern for the upper-rear leg, with the proximal femur most common. These meat-yielding parts, the femur and the scapula, are represented in equal frequencies. As discussed earlier, a scalar pattern commonly betrays a situationally conditioned selection of parts, and in this case the bias is in favor of the most meaty parts.

As was argued in the case of the small bovids, a scalar set of frequencies positively corresponding to a scalar set of utility values for the parts (see Binford 1978:23) generally betrays the cumulative result of numerous piece-butchering episodes. That is, where there is more meat than one can transport, a difficult decision is made to carry only a few parts. These parts are chosen with respect to maximizing the amount of usable food per unit of weight transported. As was the case in the argument presented with respect to the small animals, this patterning is generally most visible at kill locations because in most cases the strategic context in which this is carried out among modern hunter–gathers is one that normally includes a return to the carcass and a subsequent removal of all the usable material to the residential site. When that happens, these later acts obscure the earlier piece-butchering behavior in the overall frequencies observable at the residential site.

However, this is not the case at Klasies River Mouth. Scalar frequencies positively correlated with the utility values of the parts are demonstrable for the upper-front and -rear legs of the large bovids. This could only happen if culling took place prior to introduction of such parts to the site and the culling was severe—that is, only very select parts were transported. This extreme selectivity only applies to the adjacent segments of the upper legs since, relative to the rest of the carcass, these select parts are present only about 25% as commonly as phalanges.

Let me place this pattern in some perspective. In sites where culling has been observed, the population of parts available for selection was generally the complete animal, so that culling proceeded with respect to the relative properties of the animal as a whole. It culling was practiced, the femur would be selected over the shoulder and they both would be carried to the exclusion of the lower legs and phalanges. The result at a site where culling had occurred would be a scalar set of bone frequencies correlated with measured utility value for the parts in question. This situation is illustrated in Figure 3.20, in which the frequencies of front and rear leg-bones remaining on Nunamiut kill sites are plotted against the utility values for those same bones.

We clearly see an inverse relationship between the frequency of parts and their utility values. Stated another way, what was carried away was positively correlated with the utility values like the upper-leg parts seen at

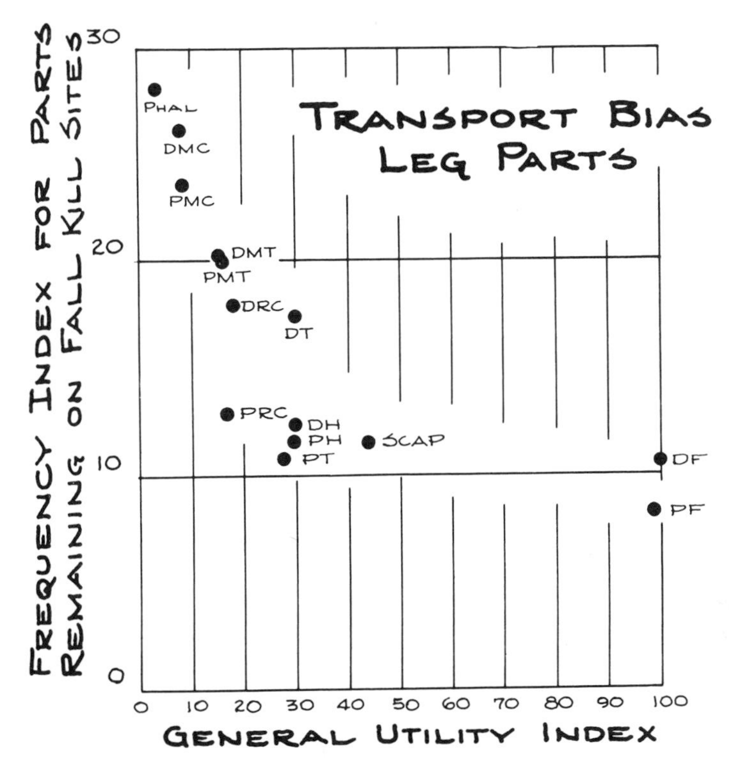

Figure 3.20 Front- and rear-leg parts of Nunamiut kill populations (Binford 1978: Table 2.8, Column 2) scaled against utility values (1978: Table 2.6, Column 10). DF, distal femur; DH, distal humerus; DMC, distal metacarpal; DMT, distal metatarsal; DRC; distal radiocubitus; DT, distal tibia; PF, proximal femur; PH, proximal humerus; PHAL, phalange; PMC, proximal metacarpal, PMT, proximal metatarsal; PRC, proximal metacarpal; PT, proximal tibia, SCAP, scapula.

Klasies River Mouth. On the other hand, in marked contrast, the frequencies of segments from the vertebral column are inversely correlated with utility values at Klasies River Mouth, as are the relative frequencies of lower-leg parts relative to upper legs. The relative frequencies of the major segments (for instance, the lower versus upper legs, or components of the vertebral column versus the head) appear to be inversely correlated with the value of the parts as potential food, whereas within high-yield segments, such as the upper legs, the relative frequencies of parts are positively correlated with potential food yield. Such a pattern—bias in favor of parts of marginal utility when viewed from the perspective of the total anatomy versus bias in favor of parts of most utility when viewed from the perspective of a particular anatomical segment of high utility—leads me to the conclusion that the animal as a whole was not the population of parts available to the hominids selecting units for return to the site at Klasies River Mouth. The manifest pattern (see Figure 3.21) is one that might be best described as "penny wise and pound foolish." I am confident that the explanation for this pattern rests with the state of the large-animal carcasses exploited by the hominids. In one sense this is the exact assemblage form previously suggested as referable to a scavenging hominid. Most often the parts remaining at a carcass are those of marginal utility, such as lower legs, and head and neck parts. These are the parts most commonly transported. Sometimes, however, the scavenger might encounter a relatively unexploited carcass from which he could obtain parts of maximum utility. Here we see a surprise, because when a large-animal carcass that had not been exhaustively exploited was encountered, it was not transported complete and then shared at the base camp. Instead, this large package of food was piece-butchered in a *gourmet* fashion: only a few of the choice parts were

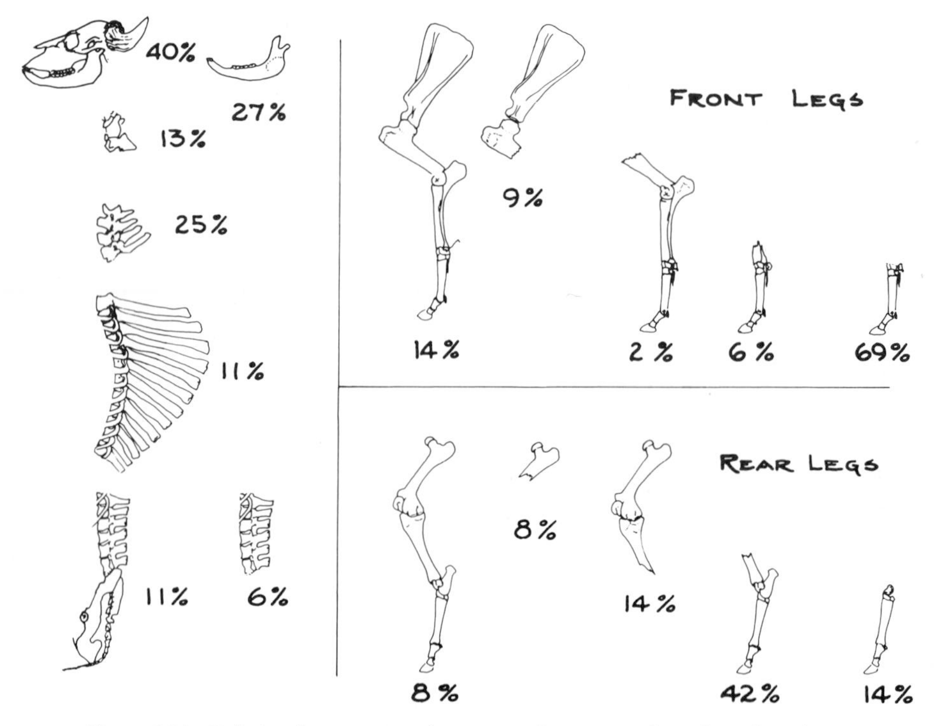

Figure 3.21 Relative frequencies of anatomical segments from large bovids introduced at Klasies River Mouth Cave 1.

taken home. I cannot imagine this type of pattern being produced by modern hunters, by whom most hunting is carried out to obtain food in packages larger than the single meal demand of the hunter. Hunters carry back much more meat than they could eat, and this large supply is then shared with other members of the group. If this had been done for the carcasses from which the high-meat-yielding parts had been obtained, they should not have been culled within the high-yielding segments. There should not be an inverse relationship between the high-yield parts of the axial skeleton such that the pelvis and lumbar vertebrae are less well represented than are the scapula and proximal femur.

The conclusion that seems inescapable is that in the data from both the small and large bovids, the most common units or package sizes of animal foods introduced to the site were quite small, and probably were selected for transport after the procurer had fed at the find location. The small size of the selected units introduced from high-yield parts (which would have been available only if the carcass was essentially unexploited by other animals) indicates that the planning depth of the hominids was very shallow; there was certainly no storage, and perhaps no planning beyond the next meal for a very small feeding unit that perhaps was no larger than one or two individuals. This small consumer-unit character to the parts introduced is supported by the biased introduction of lower-leg parts, segments that yield only a few onces of bone marrow.

It seems justifiable to infer that hominids were probably killing some small bovids but were generally consuming the choice parts at the kill (see Figure 3.18). Only rarely were complete small bovids (generally minus the lower legs) introduced into the site; the most common practice was a biased introduction of the scapula, from which the other bones with attached meat had generally been culled. In short, the segmentally transported parts were of moderate utility and were generally culled into small package sizes. Among the large animals (see Figure 3.21), the most commonly introduced parts were the marrow-yielding lower legs and other parts of marginal utility, such as the head and neck. These would be the parts most commonly available at carcasses already ravaged by other predator–scavengers. On the other hand, choice meat-yielding parts were occasionally introduced, but like the meat-yielding parts of the small bovids, these were drastically culled into parts of very small size prior to introduction to the site. If such choice parts were available, the nearly complete carcass of the animal must have been present, yet there was no attempt to remove for future use all the available food. Instead, a gourmet strategy was employed, returning only small parts of the choice segments. This clear pattern is totally inconsistent with a model of behavior that imagines considerable planning depth in the food-procurement tactics and general sharing among a band-size consumer unit. The lack of planning depth is clearly indicated. The implications for sharing among residential partners are either that (1) the size of the residential unit was very small, with only two or three individuals, or (2) sharing was not a planned practice among the members of a residential unit.

CHAPTER 4

A Pattern Recognition Study

The Klasies River Mouth fauna was selected for study because, judging from published reports by Klein (1976), it had all the provocative properties suggestive of a scavenged assemblage. A more detailed study of anatomical part frequencies, employing corrections for ravaging, provided even more evidence consistent with an interpretation of scavenging, particularly for the larger animals represented. Because it has already been suggested that a scavenged assemblage should have some general characteristics referable to the ravaged and drying state of a carcass apt to be scavenged, I now turn to the exciting task of reporting on the study of inflicted marks.

The Axial Skeleton

Some problems in studying the axial skeleton appear to have been caused by collector bias against parts that excavators thought could not be identified as to species. These included primarily parts from ribs and broken skulls, and bodies of vertebrae. If correct, this means that the estimated minimal number of elements (MNEs) should be taken with some skepticism; however, the relative frequencies of breakage, cut marks, and other modifications on the parts actually present should be fairly representative.

TABLE 4.1

Horn-Core Bases Tabulated by Body Size

Bovid class	MNE (1)	Cut marked No. (2)	Cut marked % (3)	Hack marked No. (4)	Hack marked % (5)	Gnawed No. (6)	Gnawed % (7)
I	22	0	0	0	0	0	0
II	9	1	.11	0	0	1	.11
III	18	0	0	2	.11	1	.06
IV	23	0	0	4	.17	2	.09
V	0	0	0	0	0	0	0
	72	1	.01	6	.08	4	.06

Horn

In counting MNEs I have focused on horn bases where there is some segment of adhering skull, or on horn tips where there is no ambiguity as to the unit representation of a single horn. The bases actually turned out to be the most diagnostic fragment (Table 4.1).

Two facts are provocative in this tabulation. First the small-bovid class is well represented and these are commonly unmodified and unbroken small horns; relatively few horn bases referable to the small–medium- and medium-size bovids are present. Second, and equally important, is the fact that these horn cores are rarely broken (when observed, they were almost exclusively kudu horns) and survive in the assemblage as essentially complete cores (only 0.07% were split by percussion impact). Figure 4.1 illustrates the typical state in which the horns from the moderate-body-size animals occur in the assemblage. On 2 of the 18 examples of broken horn cores there are massive hack marks at the base of the horn, obviously made with a heavy chopping instrument. This is commonly associated in my experience with the removal of horns or antlers and is unrelated to skinning activities of fresh carcasses.

In marked contrast to the horn cores of the moderate-size animals are the horn core fragments from class IV size animals, most of which are in fact *Taurotragus* (eland) horn cores. These, without exception, have been intentionally split open by heavy percussion blows, presumably to recover the small amount of pulpy tissue from the inside of the cavity near the proximal end of the horn core. Figure 4.2 illustrates the pattern of breakage and shows also the relationship of the impact blows to the pulp cavity within the horn cores. These are large horns, and the heavy blows required to split the

Figure 4.1 *Hippotragus* horn cores, illustrating their unbroken state.

Figure 4.2 *Taurotragus* horn cores, illustrating their split condition.

fresh horn cores are obvious. Consistent processing of horns for relatively small amounts of bloody pulp suggest a regular use of a very marginal food which was almost exclusively extracted from the animals in the kudu–eland size range. The extensiveness of this practice of processing horn cores for pulpy food is further emphasized by the fact that an additional 126 fragments of split-horn core were not diagnostic of a distinct element, but certainly represented a considerable pile of debris from the processing of horns by percussion techniques.

Of some interest is the fact that among these nondiagnostic fragments were three that showed distinct tooth scarring by gnawing animals, with the additional property of the breakage clearly interrupting the pattern of tooth scarring. This demonstrates that the breaking of the horn cores at least in those cases had occurred after the horns had been gnawed by nonhominid predator scavengers.

Occipital Condyles

Table 4.2 summarizes the information on the frequency of occipital condyles in the assemblage together with the information on cut marks, hacking and animal gnawing.

In my experience it has been repeatedly noted that cut marks across the occipital condyles are very common and derive from the removal of the head with cutting tools (see Binford 1981:102). It is interesting that relatively high frequencies of such marks are noted for the small animals in this assemblage, while no such marks were observed on any occipital condyles remaining from the larger animals. In similarly striking contrast is the lack

TABLE 4.2

Occipital Condyles Tabulated by Body Size

		Cut marked		*Hack marked*		*Gnawed*	
Bovid class	*MNE (1)*	*No. (2)*	*% (3)*	*No. (4)*	*% (5)*	*No. (6)*	*% (7)*
I	2	1	.50	0	0	0	0
II	9	3	.33	0	0	0	0
III	8	0	0	0	0	0	0
IV	26	0	0	0	0	2	.08
V	2	0	0	0	0	1	.50
	47	4	.09	0	0	3	.06

of animal gnawing indicated for the parts from small animals, but the presence of such gnawing on the condyles of the bovids from the two largest body-size classes. Among modern hunters, cut marks on the condyles are generally more obvious the larger the animal and are less marked the smaller the animal. Cut marks are less apparent on small animals because torque and leverage are more of a meaningful aid in removing the head from the articulation with the atlas vertebrae.

The consistent lack of cut marks on the occipital condyles of the large animals reflects a different context of dismemberment than the standard butchering suggested by the marks on the smaller animals. Equally striking is the presence of animal gnawing, indicated by tooth-scored areas across the occipital condyles (see Binford 1981:46–47 for a definition of tooth scoring). This had to occur after the head was disarticulated from the atlas vertebrae. There were no tooth-scored bones that also exhibited cut marks; therefore, no clue remains on these parts to the sequence of access to the bones for men versus animals.

Maxillary Arcs

The faunal remains were so extensively broken that the largest, most-recognizable unit of the cranium was the dental arc of maxillary teeth (Table 4.3).

Aside from the relative frequencies, the single most interesting feature was the cut marks on the fragments remaining from medium-size bovids. The marks were all identical to the marks I previously illustrated (Binford

TABLE 4.3

Maxillary Dental Arcs Tabulated by Body Size

		Cut marked		*Hack marked*		*Gnawed*	
Bovid class	*MNE (1)*	*No. (2)*	*% (3)*	*No. (4)*	*% (5)*	*No. (6)*	*% (7)*
I	22	0	0	0	0	0	0
II	14	0	0	0	0	0	0
III	9	4	.11	1	.03	0	0
IV	7	0	0	0	0	0	0
V	6	0	0	0	0	0	0
	58	4	.07	1	.02	0	0

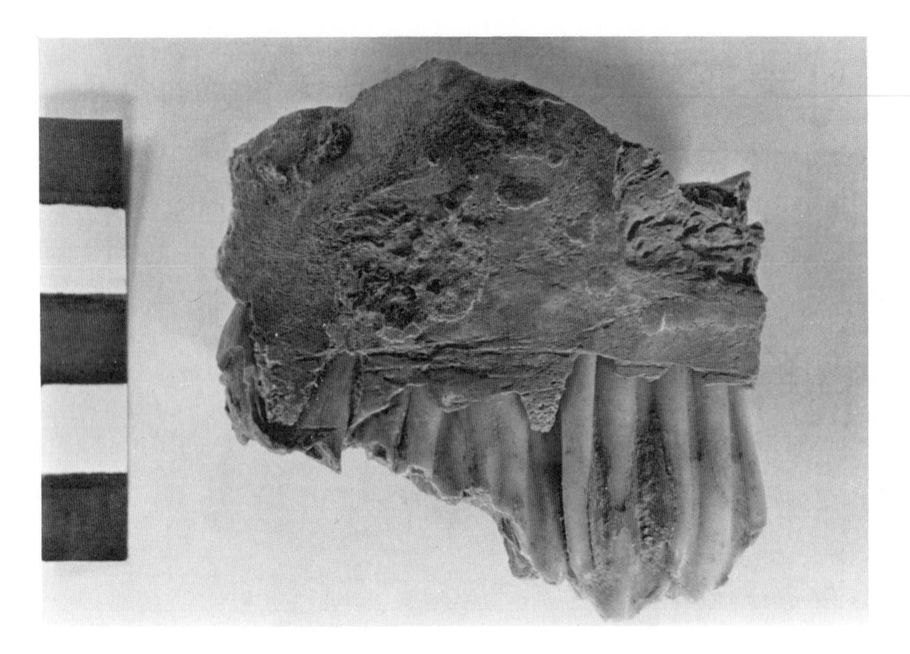

Figure 4.3 Position of cut marks along maxillary teeth.

1981:110, Figure 4.19) as exemplary of heavy-handed butchering of frozen or still carcasses:

> there is a distinctive mark left on the skull from a particular cut made during the removal of the mandible when the entire unit is frozen. Since the head with attached mandible is completely stiff, a deep and long cut is made from the insert of the masseter muscle along the upper lip area above the upper molars directly back across the ascending ramus of the mandible, severing the masseter muscle completely. Once this is done the mandible may be manipulated slightly and the task of removal is made much easier. (Binford 1981:109)

Obviously, the carcasses butchered at Klasies River were not frozen but they could well have been very stiff prior to butchering—particularly if many of the animals were scavenged rather than hunted. It is of further interest that the single example of hack marks on the maxilla is in the same location as the cut marks shown in Figure 4.3.

Teeth

As I previously noted, the units in terms of which faunal assemblages were most likely constituted were anatomical segments. After processing for

TABLE 4.4

Summary of Loose Teeth plus Those Remaining Encased in Bony Parts, Tabulated by Body Size

	Maxillae			*Mandibles*		
Bovid class	*MNE* (1)	*MAU* (2)	% (3)	*MNE* (4)	*MAU* (5)	% (6)
I	136	13	.24 (I–II)	324	29	.29 (I–II)
II	92	9		229	21	
III	237	23	.68 (III–V)	436	40	.61 (III–V)
IV	374	37		583	62	
V	87	8		238	22.5	
	839			1810		

food and during consumption in the past as well as subsequent attrition and breakage we largely recover fragments of anatomical segments actually used in the past. This is perhaps nowhere more evident than with loose teeth, which almost certainly were introduced to the site still seated in the maxillary and mandibular dental arcs of jaws and heads. Table 4.4 summarizes the MNEs (in this case, mandibles and maxillae) indicated by the loose teeth, when differences in age and differential frequencies of specific teeth are ignored. Since teeth are some of the most resistent anatomical elements to destruction, they may be frequently taken as fairly reliable estimators for the numbers of jaws and crania that had been introduced to the site.

The information from Table 4.4 may be used in several ways. However, first it is perhaps appropriate to use the data on loose teeth from the maxilla as a control in looking at the other parts recognizable from the crania. Table 4.5 summarizes the comparisons of parts of the cranium. It is clear that the occipital condyles and the maxillary teeth are essentially identical in their proportions from small to medium–large animals. On the other hand, the horn cores are considerably better represented among the small animals. This is what one would expect if meat-yielding parts were being transported to the living site; namely, the larger the animal, the greater the likelihood of abandoning the horns at or near the location of procurement (see Binford 1978:59–64). On the other hand, for very small animals that could be transported whole with ease, horns would be normally introduced. As has already been noted, the horns of medium and particularly medium–large animals have been processed for recovering an edible pulp from the core cavity, whereas the small animal horn cores have not been so processed. This must most certainly be a very marginal food at best. Consequently, we would expect that regular processing for such a food would be done only when better foods were in low supply. This is analogous to the

TABLE 4.5

Frequency Comparison among Parts of the Cranium

	Horn		Occipital condyles		Maxillary teeth		Maxillary arcs	
Bovid class	*MAU* (1)	% (2)	*MAU* (3)	% (4)	*MAU* (5)	% (6)	*MAU* (7)	% (8)
I and II	15.5	.43	5.5	.23	22	.24	18	.62
III–V	20.5	.57	18	.77	68	.76	11	.38
	36		23.5		90			

processing of mandibles by the Eskimo for a small and nutritionally marginal bit of pulp from the base of the roots of the mandibular teeth (Binford 1978:23–32). The fact that horns are processed from relatively large animals (kudu and eland, primarily) is consistent with the view that the large amount of meat from these animals was not available to the consumers of the pulpy contents of the horn cores. On the other hand, the lack of horn processing for the smaller animals is consistent with the possibility that better foods were available—the meat of these animals. I am suggesting that at least some crania with attached horns were scavenged from the death sites of moderately large animals and were introduced to the site, where the horns were processed for a very marginal food.

The nearly reversed frequencies for maxillary arcs relative to the frequencies of maxillary teeth among small and medium–large forms is most reasonable seen as reflecting differential breakage.

Mandible

The mandible is one of the most common parts in the Klasies assemblage. In fact, for the medium (class III) through the large (class V) animals it is absolutely the most common anatomical part. Table 4.6 summarizes the relative frequencies of mandibular and maxillary elements as indicated by tooth counts. This comparison is taken as an indication of the relative frequencies of mandibles versus crania introduced to the site.

The ratios shown in Column 3 of Table 4.6 tell the story nicely. Among the two smallest body-size classes (I and II) the numbers of crania are 45 and 43%, respectively, as common as are mandibles. This situation (bias in favor of mandibles) was noted repeatedly among the Nunamiut Eskimo

TABLE 4.6

Comparisons of MAUs Indicated by Teeth from the Maxilla and Mandible

Bovid class	*Maxillary teeth MAU (1)*	*Mandibular teeth MAU (2)*	*Column 1 / Column 2 (3)*
I	13	29	.45
II	9	21	.43
III	23	40	.58
IV	37	62	.60
V	8.5	22.5	.38

hunters, where mandibles were introduced to sites from fresh kills in numbers exceeding their expected frequencies, given objective measures of their food value (see Binford 1978:199). The overall bias in favor of mandibles in the Klasies data might be understood in similar terms, but this does not help in understanding why there is a shift in the relative maxilla/mandibular ratio, seemingly related to body size as illustrated in Table 4.6.

In data from animal kills and dens it has been noted that from relatively small prey animals, mandibles tended to be far more common in animal dens than parts of the cranium, presumably because with very small animals the cranium was commonly destroyed as a unit at the time predators originally fed on small-prey animals (see Binford 1981:232–233; Richardson 1980b). On the other hand, at animal dens the body-size range from medium to medium–large bovids exhibits a marked increase in the frequency of crania relative to mandibles, although mandibles continue to be absolutely the most common part of the head represented.

Following this comparison still further, it is noted that for prey animals over 85 kg in body weight, parts of the skull were equal to or slightly greater than the frequency of mandibles in den assemblages (presumably hyaenas were the denning animals in this example). A proportional increase was also noted for wolves (see Binford 1981:Table 5.01, Column 28; and Table 5.03, Columns 4–6). On the other hand, as body size increased beyond 80 to 85 kg, the overall number of head parts decreased relative to lower-leg parts at animal dens. This fact suggests to me that, for both hyaena and wolves, the head parts from relatively large prey are most often introduced from small individuals (young or immature). Figure 4.4 compares the percentage MAU for animals of different body size from both the Swartzklip (Binford 1981:216) and Bent Creek (Binford 1981:213) animal-den assemblages with the same values for the body size classes represented at Klasies River Mouth. It can be seen that there is a striking parallel between

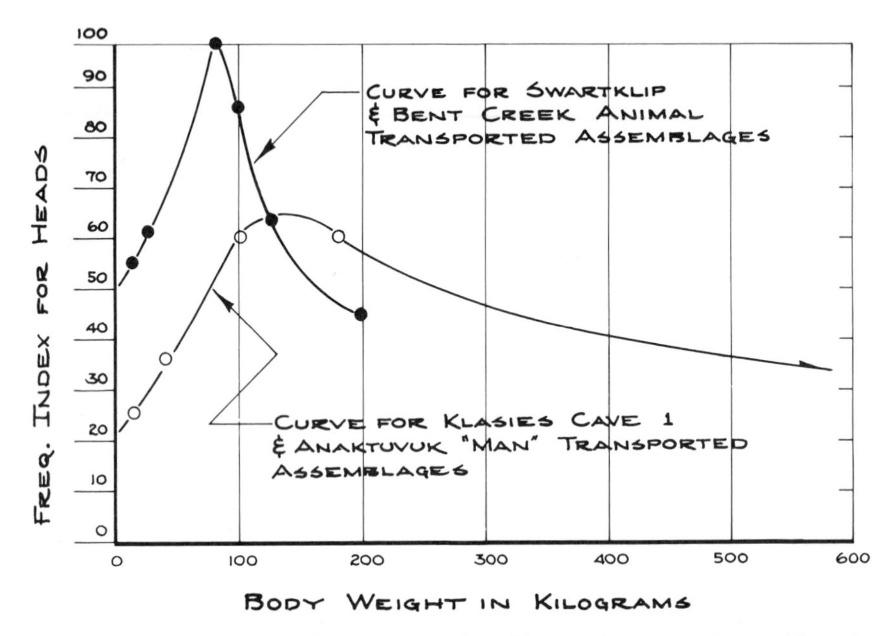

Figure 4.4 Comparison between animal- and hominid-transported assemblages in the relationship between body size and frequency of head transport.

the two data sets. The human curve is lower and covers a much greater range of body sizes, and it peaks at a greater body size (130 kg). Nevertheless, the overall character of the curve is strikingly similar.

This comparison illustrates nicely that at least in the transport of head parts back to living sites, the occupants of the rockshelter at Klasies River Mouth behaved in ways directly analogous to other predator–scavengers transporting head parts back to their dens. Such a provocative comparison is further amplified when it is noted that the proportions of mandibles to crania for size classes III and IV approach those abondoned at predator kill sites for prey animals in the same body-size range (see Binford 1981: Table 5.02, Column 2). All these observations are consistent with the view that the removal of head parts from kill–death locations by occupants of Klasies was conducted in essentially the same context as were the removals of parts to dens by nonhominid predator scavengers.

Whatever the direct conditioning factors standing behind these patterns, it is clear that the size of the animal's head is a major determinent of the frequency with which it is transported. This means that if size is the major conditioner, as it appears to be, then this factor will condition the removal of heads even within species, since individuals vary in head size as a function of age and sex. If the transport filter is size, then we can expect

TABLE 4.7

Modifications to Mandibles, Tabulated by Body Size

		Cut marked		*Hack marked*		*Gnawed*		*Dentary*	*Breakage*
Bovid class	*MNE (1)*	*No. (2)*	*% (3)*	*No. (4)*	*% (5)*	*No. (6)*	*% (7)*	*no. (8)*[a]	*no. (9)*[b]
I	30	2	.07	0	0	0	0	29	0
II	27	1	.04	0	0	0	0	22	0
III	25	4	.16	0	0	0	0	11	7
IV	32	2	.06	3	.09	3	.09	6	6
V	21	1	.05	5	.24	2	.10	9	9
	135	10	.07	8	.06	5	.04	77	22

[a] This column indicates the number of the MNE represented by the dentary area (half mandibles).

[b] This column shows the number of dentaries with the lower margin of the mandible broken away.

differing age and sex profiles to characterize a population for head parts ranged across species of different body size. This means that the age and sex profiles of head parts occurring in a *transported* assemblage do not reflect the mortality pattern (catastrophic or attritional [Klein 1981:61; 1982]), but instead the biased transport of different package sizes away from kill–death locations.

As was the case for processed horn cores, we noted that the processing of mandibles for the truly marginal food available at the roots of the mandibular teeth was exclusively restricted to the medium-to-large animals (body-size classes III to V). The mandibles of the smaller animals are never broken along the ventral margins of the horizontal mandibular ramus. Equally similar to the situation with horn cores, it could be argued that with the small animals the quantity of this marginal food is so limited that recovery is not worth the effort. On the other hand, if the medium-to-large animals were being hunted, they would supply large quantities of very high quality food and it is reasonable to wonder why, given so much food, so much effort was expended to recover this truly marginal morsal? As was the case with the horn cores, there is further evidence that there was very little high-quality food available when the large animal parts were processed.

Several lines of evidence support this view. In Table 4.7 it is noted that cut marks regularly occur on mandibles of both large and small animals (.05% of size classes I and II, and .09% of MNE in size classes III–V). What is more interesting are the differences in the kinds of cut marks occurring on the small versus the large animals. Figure 4.5 illustrates the placements of

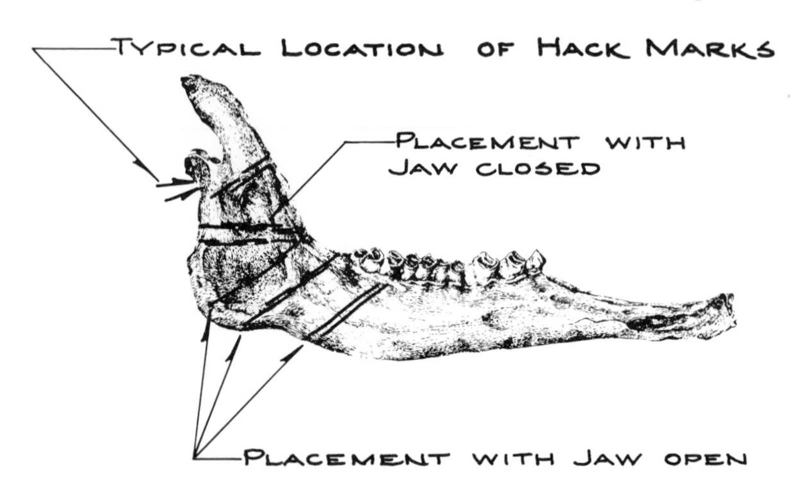

Figure 4.5 Dismemberment-mark placement when the mandible is either open or closed.

cut marks noted on the mandibles from Klasies River Mouth. It should be clear that there are basically two orientations to the cut marks: (1) oblique across the horizontal ramus, with a tendency for a concentration of such marks on the ventral margins of the masseteric fossa, or (2) diagonally across the ascending ramus just below the mandibular condyle. All these oblique marks are generally inflicted when the mouth is held open during butchering. This is most commonly possible when a fresh, supple carcass is being cut up. On the other hand, when a dry or stiff carcass is being butchered, the mouth is shut and it is almost impossible to open it. When this is the case, cut marks such as the horizontal ones indicated in Figure 4.5 are commonly made running across the ramus, impacting the maxillary arc just above the upper third molar (as shown in Figure 4.3). This is most common when the carcass is stiff but not yet dry. When a carcass is dry, the use of sharp cutting tools, even modern steel knives, is ineffective on the dry and desiccated skin, muscle, and tendon. When a dry carcass is being dismembered, chopping and hacking and heavy handed "sawing" with breaking blows is the appropriate procedure. In addition, I have noted that when experimentally butchering carcasses, if the carcass is supple, sharp incising cuts with unretouched flakes and blades are very effective. When this is done, single small V-shaped cut marks are common, running some distance across a bone (only interrupted by changes in surface shape of the bone). If retouched flake knives are being used, a similar pattern may be seen, but there may be "hairline" parallel sets of marks, representing the impacts on bone of small peaks or unaligned high points on the cutting edge. On the other hand, when relatively stiff (and frozen) meat and tendon are being addressed, the small sharp cutting implements are ineffective. However,

Figure 4.6 Hack marks on the base of the mandibular condyle (*Taurotragus*).

with a large hefty edge—as on a handax, core scraper, or other large biface—one uses a combination of hacking and sawing motions. These leave whole sets of short and frequently thick cut marks, many of which are not exactly parallel and may be separated from one another by several millimeters. This irregularity arises from the twisting and slight reorientation of the edge to the bone surface as the large tool is sawed and bruised into the resistant material. This type of clustered and larger scarring is well illustrated in Figure 4.5.

Hack marks from such treatment were noted on mandibular condyles of eight of the mandibles from medium–large (eland-size) and large bovids (size classes IV and V). All the hack marks and the horizontal cut-marks illustrated in Figure 4.6 were observed on mandibles of the three largest body-size classes (III–V). This is very strong evidence that the latter were dismembered by hominids after the carcasses had become stiff and dry, not when they were fresh and supple.

In addition, it was noted in all cases where the cut marks were parallel to the tooth rows, suggestive of butchering when the jaw was closed and hence rigid, the cut marks were of the grouped, short, "set" variety and were not the long marks shown in Figure 4.7. These contrasts are taken to

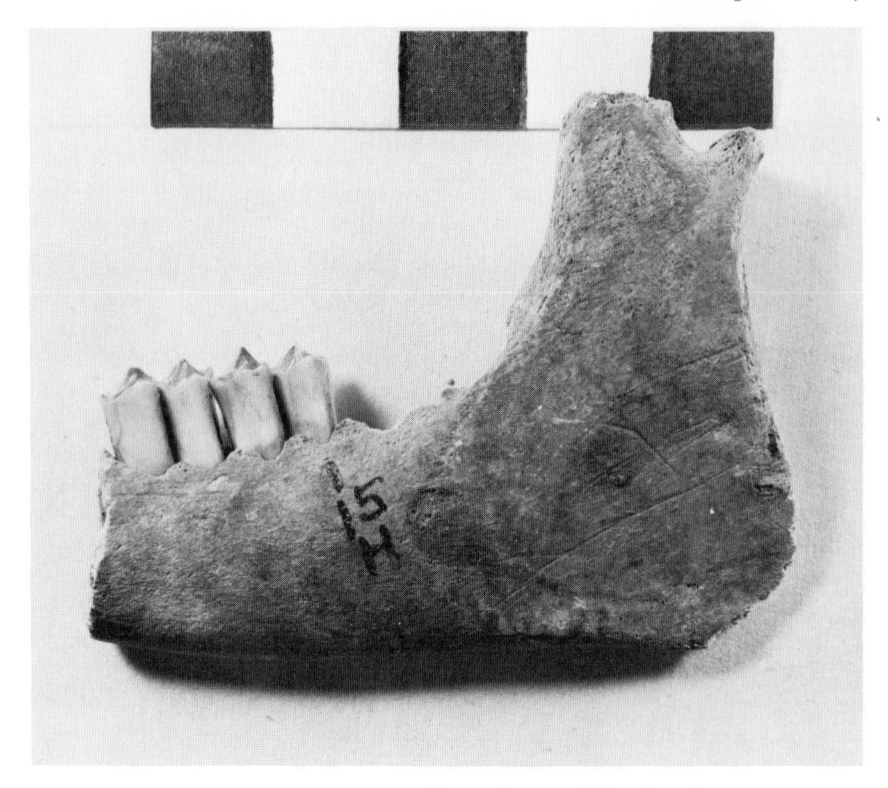

Figure 4.7 Open-mount cut marks on the mandible of *Raphicerus.*

indicate the use of different tools when butchering fresh and supple carcasses versus rigid and presumably partially desiccated ones. Sharp cutting edges, such as occur on freshly struck flakes, appear to be the primary instruments when the carcasses are fresh. When rigid, a heavy tool that can be used for hacking, bruising, and sawing seems to have been used. Something like a handax or edge of a thick core is a likely candidate.

As in the case of other head parts, all the animal gnawing noted on mandibles was referable to bones of animals in the large-size classes. This observation further supports the view that nonhominid scavengers had ravaged the carcasses of at least some of the larger animals.

Vertebral Column

Cervical Vertebrae

It is interesting that the proportions of atlas to axis (.64%; Table 4.8) are roughly the same for both small and large animals. On the other hand,

TABLE 4.8

Atlas and Axis Vertebrae, Tabulated by Body Size

Bovid class	MNE (1)	Cut marked No. (2)	Cut marked % (3)	Hack marked No. (4)	Hack marked % (5)	Gnawed No. (6)	Gnawed % (7)
			ATLAS				
I	7	3	.43	0	0	0	0
II	5	1	.20	0	0	0	0
III	4	0	0	0	0	0	0
IV	7	0	0	0	0	0	0
V	4	0	0	0	0	0	0
	27	4	.15	0	0	0	0
			AXIS				
I	12	1	.08	0	0	0	0
II	7	0	0	0	0	0	0
III	6	0	0	0	0	0	0
IV	13	0	0	0	0	0	0
V	4	0	0	0	0	0	0
	42	1	.02	0	0	0	0

the other cervical vertebrae are more common from animals of size classes II and III (see Table 3.5). There the atlas and axis represent only 29% of the cervical vertebrae. In the former case, numbers of skulls are being introduced, presumably with attached atlas and axis vertebrae but unaccompanied by the remainder of the neck.

Cut marks were really only meaningfully noted for the small animals, supporting the data from the occipital condyles, that only among the smaller forms was the head regularly cut from the neck. Hacking was not exhibited on the neck parts. Gnawing was noted on 8% of the cervical vertebrae of size class IV and 20% of size class V.

Thoracic Vertebrae

Because very few complete vertebrae were observed, the thoracic spines are described independently of the fragments of vertebral body (Table 4.9).

In my experience with butchering marks, one of the most consistently represented is placed roughly parallel to the orientation of the vertebral column along the base of the dorsal spines (see Binford 1981: Figure 4.21). These are produced during the removal of the tenderloin muscles, which lie in a long bundle along the vertebrae on either side of the spinous processes

TABLE 4.9

Thoracic Vertebrae: Spines Only

Bovid class	*MNE*	*Cut marked*		*Hack marked*		*Front breakage*		*Gnawed*		*Broken and Gnawed*	
		No.	*%*	*No.*	*%*	*No.*	*%*	*No.*	*%*	*No.*	*%*
	(1)	*(2)*	*(3)*	*(4)*	*(5)*	*(6)*	*(7)*	*(8)*	*(9)*	*(10)*	*(11)*
I	44	16	.36	0	0	0	0	0	0	0	0
II	43	12	.28	0	0	2	.05	0	0	2	.05
III	43	5	.12	0	0	4	.09	0	0	4	.09
IV	16	2	.13	1	.06	6	.38	1	.06	7	.44
V	10	1	.10	2	.20	6	.60	2	.20	8	.80
	156	36	.23	3	.02	18	.12	3	.02	21	.13

(this removal is well illustrated in Binford 1978: Figure 2.1). In my control data from the Nunamiut Eskimo, over 50% of the thoracic vertebrae are marked along the base of the thoracic spines (see Binford 1981:Table 4.02).

Table 4.9 demonstrates that, like the Nunamiut data, a relatively large number of thoracic spines are scarred by cut marks at Klasies River. Of even greater importance is the very high frequency of spines from animals in the two smallest body-size classes (size classes I and II), in which 36 and 28%, respectively, of the identified spines were marked. In contrast only 12, 13, and 10%, respectively, of the spines from larger body-size animals exhibited cut marks. This contrast provides further evidence of more dismemberment by cutting among the smaller animals.

In my experience with field-butchering situations, the larger the animal, the greater the degree of dismemberment in the field to facilitate transport. For instance, when I have been with hunting parties on foot or even with pack dogs and an animal the size of a moose (408–680 kg) was killed, field dismemberment included the disarticulation between the scapula and proximal humerus, as well as between the distal radius and the proximal metacarpal. On the other hand, when caribou were field-butchered, the front leg was commonly disarticulated only between the distal radius and the proximal metacarpal. These differences reflect that more-extensive field butchering of large animals makes possible the transport of "reasonable"-size units when the prey is very large. This same principle applies to the axial skeleton. For instance, in field-butchering caribou, it is common to remove the complete vertebral column as a unit, whereas in all the cases I observed of butchering animals of size of moose, the vertebrae were butchered into at least four units: neck; thoracic; lumbar, with the sacrum remaining attached to the lumbar vertebrae; and pelvic units, with the pelvis frequently separated into two units, a right and left side.

This experience leads me to expect greater numbers of cut marks on a greater number of bones from large animals prior to the introduction of parts to a site of consumption. We observe the reverse at Klasies River. In the comparison made here, cut marks have been more common on the small animals and less common on the bones from larger animals. I think the conclusion is inescapable that *knife butchering* was less common, the larger the animal. It is unlikely that this would be the case if large animals were being field butchered after being killed by man, or when their carcasses were still fresh.

This interpretation is further supported in the case of thoracic remains by a number of additional provocative and informative traces remaining on the bones. Very provocative is a property that I noted while studying wolf kills; namely, that wolves tend also to go selectively after the tenderloin, but instead of using knives they sink their teeth into the muscle on either side of the thoracic spine, vice down, and pull back. This results in punctures into the spinous process, coupled with pulling upward or away from the vertebral column. The result of this action is the frequent breaking away of the dorsal edge of the spinous process, frequently coupled with tooth puncture or pitting marks along the ridge of the thoracic spine. Figure 4.8 shows a

Figure 4.8 Canid feeding on a section of lumbar vertebrae, showing the "pulling up" action of removal of the tenderloin.

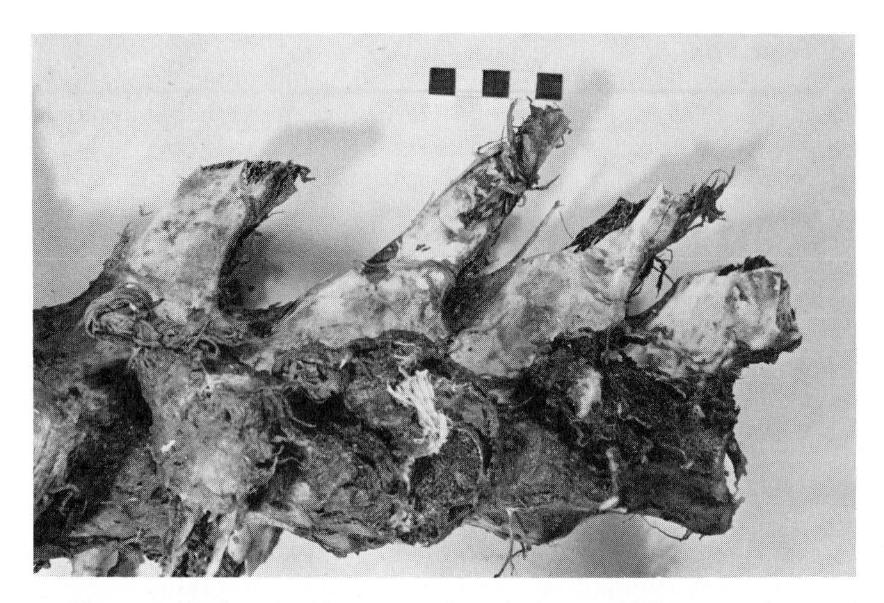

Figure 4.9 Modern eland (*Taurotragus*) vertebrae gnawed by hyaena in the Nossob Valley, South Africa.

canid holding down a section of lumbar vertebrae and pulling up along the lumbar dorsal spines. Figure 4.9 illustrates the type of breakage resulting, as observed on thoracic vertebrae of an eland scavenged by hyaenas. This same condition is well illustrated by the dried carcass of a blue wildebeest that had been scavenged by hyaenas (Figure 4.10).

This type of breakage, parallel to the vertebral column and localized on the dorsal ridge of the spine, is tabulated by body size in Columns 5 and 6 of Table 4.9. It is clear that, unlike cut marks from knifelike tools, this breakage pattern, diagnostic of carnivore feeding, is most common on the bones of large animals and is absent on the thoracic spines of small animals. The frequency of this form of breakage is paralled by gnaw marks (similar to those illustrated in Binford 1981: Figure 3.32) produced by animals and is reinforced by the relatively high frequencies of gnawing indicated on the centrum or body fragments (Figure 4.11) of thoracic vertebrae (Table 4.10). Twenty percent of the thoracic body fragments showed evidence of gnawing in the form of tooth punctures, tooth scoring, or crenulated edges of various processes from the largest two body-size classes. No such gnawing was exhibited on the vertebrae from the small-body-size animals. This pattern is amplified still further by the evidence of hack marks inflicted by heavy hacking–chopping actions. As has already been suggested, hacking and

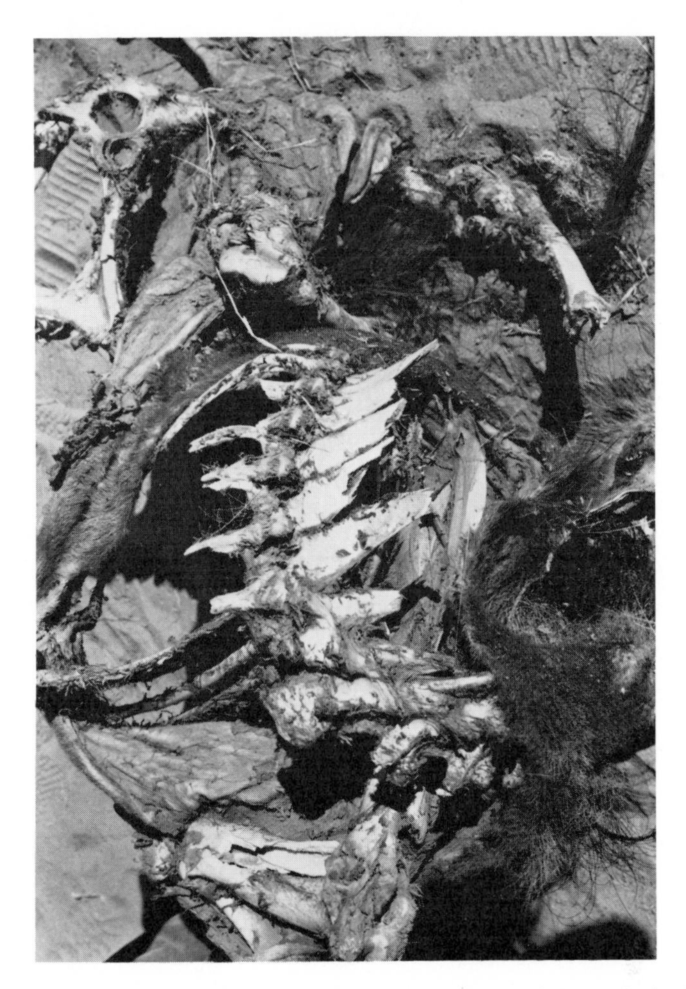

Figure 4.10 Modern wildebeest (*Connochaetes*) carcasses fed upon by both hyaena and jackals, showing upper breakage on the dorsal spines and neck breakage of the ribs (Nossob Valley, South Africa).

chopping are essentially the only ways to dismember a dry or rigid carcass. While it is possible to dismember a fresh carcass with a sharp cleaver, a dull bruising chopper is almost impossible to use in going through heavy muscle and tendon that are fresh. Judging from the character of the hack marks, they were inflicted when the carcass was relatively dry and stiff. These facts all fit together to support the inference that, in the main, carcasses of the larger animals were not fresh, and had been probably fed upon by non-

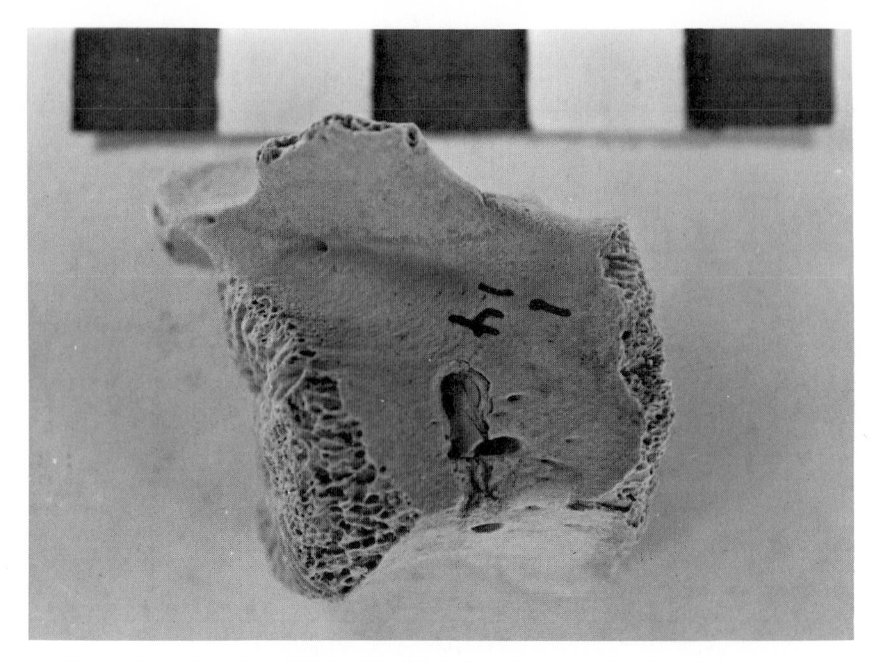

Figure 4.11 Vertebra of *Pelea* (Vaalrhebok), showing animal-tooth punctures.

hominid predator–scavengers prior to the dismemberment and transport of usable parts by the hominids back to the site at Klasies River Mouth.

The objection could be raised that the feeding and gnawing of bones by nonhominid carnivore feeders could have been done on site at Klasies River after the hominids had abandoned their living place. This view could be coupled with the denial that the hacking had to be done at the kill and could

TABLE 4.10

Thoracic Vertebrae: Centrum Only

		Cut marked		*Hack marked*		*Gnawed*	
Bovid class	*MNE* *(1)*	*No.* *(2)*	*%* *(3)*	*No.* *(4)*	*%* *(5)*	*No.* *(6)*	*%* *(7)*
I	64	1	.02	0	0	0	0
II	69	0	0	0	0	3	.04
III	89	0	0	0	0	8	.09
IV	54	0	0	0	0	11	.20
V	10	0	0	2	.20	2	.20
	286	1	.01			24	.08

represent a processing alternative of transported parts that had become stiff and partially dried out while awaiting consumption in the living site itself.

This is, of course, the position generally taken by Glynn Isaac (1971:288) and H. Bunn (1982:495), as well as by Mary Leakey (1971:43) with regard to the evidence of carnivore gnawing on the bones from the famous sites at Olduvia Gorge and, more recently, at Koobi Fora. I will have more to say about this possibility as the argument from the Klasies River Mouth bone collection progresses. Obviously this study was done with this problem in mind.

Ribs

During the course of my studies of hunting, butchering, and consumption of animal products by living peoples, I learned a number of things about breakage morphology and use of ribs. Most commonly the ribs are removed during initial field butchering as a unit—a rib slab—and this is accomplished by breaking the rib unit back or up against the vertebrae (Figure 4.12) and then by cutting along the ventral surface of the broken

Figure 4.12 Nunamiut Eskimo removing a rib slab by cracking it back against the vertebrae. This leaves a distinctive fracture pattern.

ribs to free the slab from the vertebrae. This results in a characteristic break coupled with distinctive cut or slicing marks on the ventral surfaces of ribs. When the carcass is fresh, the ribs crack back, so that the stress is primarily focused on the neck of the rib; that is, on the short section of bone between the head and the tubercle at the articular end of the rib. This breakage results in the tiny head's remaining attached to the thoracic vertebrae, and the break just forward of the tubercle therefore characterizes the rib as removed from the carcass.

With very large animals this removal of a slab unit is too difficult, and ribs may be removed in units of three or sometimes even broken back one at a time. When the latter is the case, there is a torque set up relative to the remaining, unbroken ribs and the break tends to be a very distinctive diagonal break just back of the tubercle, across the costal groove. In both cases, the breaks are correlated with slicing cut marks on the ventral surface of the rib adjacent to the break. On the other hand, when the animal is dry and the tendons attaching the ribs to the vertebrae are desiccated, the proximal end of the rib is immobile in a vicelike grip of dry tissue. If one wishes to remove ribs under these conditions, they must be broken and chopped through roughly across the shaft of the rib, just distal to the angle of the rib, since all the attachments to the vertebrae are in the area between the angle and the head of the rib. This results in a complete rib head's breaking off just distal to the angle and remaining attached to the thoracic vertebrae, with the dismembered rib unit itself having no anatomical segments of the head attached.

Table 4.11 summarizes the breakage pattern associated with the proximal ribs, as well as a tabulation of other rib segments noted in the Klasies assemblage. What is very clear is that proximal ribs with complete heads (indicative of breakage while dry), and broken across the angle (Columns 2 and 3), are progressively more common the larger the animal, whereas heads broken between the head and the tubercle (indicative of fresh carcasses dismemberment) are more frequent the smaller the body size.

This observation supports the growing body of contrasts demonstrating that the large animals consistently show evidence of having been dismembered when they were less fresh and relatively stiff. This inference is further supported by the general lack of rib fragments from the larger animals. It was noted among the Eskimo (Binford 1978:151–152) that when ribs were fresh it was common to break them open and to suck out the small amounts of pulp from the interior. On the other hand, when ribs were dry, this material tended to be putrid. Much care was then taken not to break open rib shafts during the removal of dry meat or during the boiling of meat on the bones.

Rib heads account for 61% of all the rib fragments from the three

TABLE 4.11

Breakage of Proximal Ribs and Other Rib Segments

		Complete ribs		*Broken ribs*		*Midshaft sections*		*Distal ends*	
Bovid class	*MNE (1)*	*No. (2)*	*% (3)*	*No. (4)*	*% (5)*	*No. (6)*	*% (7)*	*No. (8)*	*% (9)*
I	22	10	.45	12	.55	25	.68	18	.82
II	25	19	.76	6	.24	7	.19	16	.64
III	12	8	.61	4	.33	3	.08	11	.92
IV	8	6	.75	2	.25	1	.03	0	0
V	7	6	.86	1	.14	1	.03	1	.14
	74					37	.50	46	

largest body-size classes (III–V), suggesting that these were most likely introduced attached to the vertebrae rather than as separate parts for consumption. Viewed in another way, there are very few sections of rib shaft, suggesting (1) few complete ribs were introduced, and, if they were, then (2) they were not broken up for consumption (possibly dry and putrid inside). On the other hand, fragments of ribs from the shaft and distal ends were more common from animals in the two smallest body-size classes (I and II), so that rib heads only amount to .37% of all the rib fragments. This difference must be meaningful because the recovery efficiency for larger bones would be expected to be high and small bones low, given the large mesh screens that the excavators are reported to have used.

The story of the ribs is amplified when the data on cut marks, hacking, and gnawing are considered (Table 4.12). Hack marks are more common

TABLE 4.12

Ribs: Cut, Hack, and Gnaw Marks

		Cut marked		*Hack marked*		*Gnawed*	
Bovid class	*MNE (1)*	*No. (2)*	*% (3)*	*No. (4)*	*% (5)*	*No. (6)*	*% (7)*
I	22	0	0	0	0	0	0
II	25	0	0	0	0	0	0
III	12	2	.17	3	.25	1	.08
IV	8	3	.38	6	.75	1	.13
V	7	4	.57	4	.57	4	.57
	58	9	.16	13	.22	6	10.0

TABLE 4.13

Lumbar Vertebrae and Sacrum: Cut, Hack, and Gnaw Marks

		Cut marked		*Hack marked*		*Gnawed*	
Bovid class	*MNE (1)*	*No. (2)*	*% (3)*	*No. (4)*	*% (5)*	*No. (6)*	*% (7)*
			LUMBAR VERTEBRAE				
I	42	6	.14	0	0	0	0
II	42	9	.21	0	0	3	.07
III	46	0	0	0	0	0	0
IV	55	1	.02	0	0	16	.29
V	15	0	0	0	0	5	.33
	200	4	.02	0	0	24	.12
			SACRA				
I	8	3	.38	0	0	0	0
II	9	2	.22	0	0	0	0
III	4	0	0	1	.25	2	.50
IV	4	0	0	0	0	1	.25
V	2	2	1.00	1	.50	2	1.00
	27	7	.26	2	.29	5	.19

than cut marks, and the former are restricted to animals of the three largest body-size classes. As we have noted, hacking and animal gnawing are correlated and both are seen as indicative of the carcasses having been scavenged and rigid prior to exploitation by the hominids for transport to their site. This pattern is very robust among the rib fragments.

Lumbar Vertebrae and Sacrum

Table 4.13 summarizes the information for both lumbar vertebrae and sacra. These compromise one of the first body parts to exhibit little clear patterning regarding the placement and relative frequencies of inflicted marks relative to body size. Gnawing, on the other hand, shows a pattern of relatively high frequencies (in this case even higher than normal) on the bones from the two largest body-size classes.

Pelvis

Cut marks on the pelvis parts from the animals of medium and large size were almost exclusively across the pubis or along the symphysis. These

TABLE 4.14

Pelvic Parts: Cut, Hack, and Gnaw Marks

		Cut marked		*Hack marked*		*Gnawed*	
Bovid class	*MNE (1)*	*No. (2)*	*% (3)*	*No. (4)*	*% (5)*	*No. (6)*	*% (7)*
I	21	7	.33	0	0	0	0
II	23	1	.17	0	0	0	0
III	19	3	.16	0	0	0	0
IV	17	3	.18	0	0	6	.35
V	6	1	.17	1	.17	5	.83
	86	15	.17	1	.01	11	.13

marks are sometimes associated with the removal of meat rather than dismemberment, but may also be related to the dismemberment of the pelvis itself into right and left halves. When the latter is the case, the cuts are generally rather robust.

As Table 4.14 demonstrates, there are qualitative differences between the cut marks noted on the small-size classes of bovids. In addition, there are twice as many marks on bones from the small bovid class. This may reflect a greater number of animals butchered when they were fresh, a condition that we will see is also indicated by the types of marks present.

Certainly of great interest are the inflated frequencies for animal-gnawed pelvic bones noted among the large and very large bovids, while such gnawing is absent on the bones of medium- and smaller-size animals. With regard to the question of whether the gnawing occurred prior to introduction of the parts to the site, or whether animals scavenged debris from human consumption after the humans had left the site, there is only one bone, a fragment of ischium, that has both cut marks from stone tools (across the body of the ischium) and tooth marks from animal gnawing (along the border between the ischial ramus and the ischial tuberosity). There is no clue in this case as to which was made first, the tool marks or the animal gnawing.

The Appendicular Skeleton

Upper-Limb Bones

I treat the upper heavy-muscle-mass bones of the front and rear quarters as a set in this description.

TABLE 4.15

Proximal Femur: Cut, Hack, and Gnaw Marks

Bovid class	MNE (1)	Cut marked No. (2)	Cut marked % (3)	Hack marked No. (4)	Hack marked % (5)	Gnawed No. (6)	Gnawed % (7)
I	18	4	.22	0	0	0	0
II	11	1	.09	0	0	1	.09
III	15	0	0	0	0	2	.13
IV	25	4	.16	0	0	7	.28
V	5	0	0	0	0	1	.20
	74	9	.12	0	0	11	.15

PROXIMAL FEMUR

Among almost all the animals larger than the small bovids, the proximal femur is represented by the femoral head broken through the femoral neck so that segments of the greater and lesser trochanter are commonly absent. In my experience this is a relatively rare form of breakage, and is referable to acts of dismemberment rather than to breakage of a femur for bone marrow. This type of breakage is almost exclusively associated with the dismemberment of animals that were frozen or animals that for one reason or another, had not been butchered when they were still supple.

Most human butchers dislocate the femoral joint by placing the foot in the crotch of the animal and simultaneously pulling up and twisting the animal's leg. I have observed this as a regular procedure on animals as large as a North American moose. This is done before the animal is skinned, and forces the femoral head out of the acetabulum so that later, when the rear quarter is being removed, all that is required is to cut the connective tissue between the femur and the pelvis without having to gouge into or around the acetabulum to sever the attachment of the femoral head inside the acetabulum at the *fovea capitis*. If this procedure is not followed, cut marks around the lip of the acetabulum are almost always present (see Binford 1981: Figure 4.22, marks PS-7 and 9). If the joint is dried and essentially immobile, the most common technique is to cut through the tissue and lever the joint as if seeking to dislocate it, resulting in a fracture through the neck of the femur. This is the pattern of breakage most common among the large animals recorded in Table 4.15. Severing the rear quarter when the femur has been dislocated is correlated with diagonal cut marks across the lateral face of the body of the ilium, as well as analogous marks across the body of the ischium (see Figure 4.13). As noted earlier, these are exclusively the types of marks observed on the pelvic parts from the small animals. Such

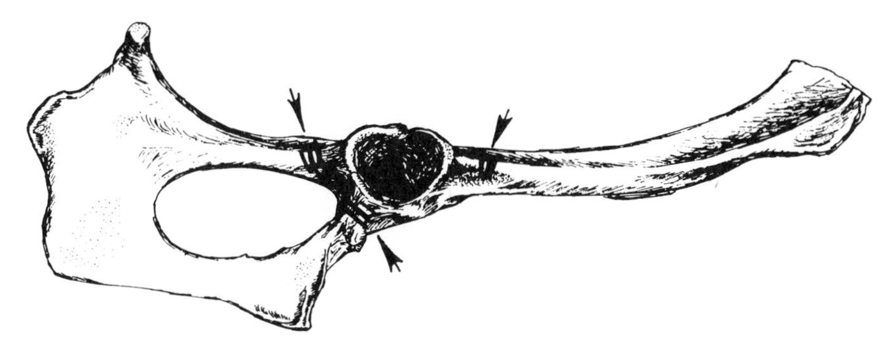

Figure 4.13 Marks inflicted on the pelvis when the femur is dislocated.

marks are generally coupled with marks across the greater trochanter or the femoral head. All the cut marks on the proximal femur of the smaller animals (size classes I and II) were of this type (these are illustrated in Binford 1981:117, mark Fp-5).

On the other hand, *all* the cut marks observed on proximal femora from animals in size class IV were marks across the neck of the femur, which is most commonly produced when the ball joint has not been previously dislocated. Such marks are frequently produced prior to levering a relatively immobile joint, when twisted breakage occurs across the neck of the femur as previously noted. At Klasies, all cuts across the neck were also associated with broken necks. This pattern clearly points to the butchering of larger animals when the joints had not been dislocated and were relatively stiff at the time of butchering.

In the case of the proximal femur, the pattern previously noted of the nearly exclusive presence of gnawing marks on bones from the animals in the larger body-size classes continues. This further strengthens the interpretation that, at the time of butchering with tools, the carcasses of the larger animals were relatively stiff and had already been fed upon by carnivores.

Breakage of the distal femur was almost exclusively through the shaft just proximal to the articular end, as is common among large and small animals, and marks were all of the same type—transverse marks across the posterior face just above the condyles (see Binford 1981:117, mark Fd-1). These marks are inflicted during dismemberment (Table 4.16).

TIBIA

The pattern noted for the femur continues with the tibia; that is, cut marks are generally present on bones of both the large and small animals, hack marks are normally absent, and gnawing is restricted to the bones of

TABLE 4.16

Distal Femur: Cut, Hack, and Gnaw Marks

		Cut marked		*Hack marked*		*Gnawed*	
Bovid class	*MNE* (1)	*No.* (2)	*%* (3)	*No.* (4)	*%* (5)	*No.* (6)	*%* (7)
I	37	6	.16	0	0	0	0
II	40	2	.05	0	0	0	0
III	26	0	0	0	0	1	.04
IV	27	7	.27	0	0	4	.15
V	9	4	.44	0	0	1	.11
	139	19	.14	0	0	6	.04

larger animals (Tables 4.17 and 4.18). Unlike the femur, the kinds of cut marks are essentially the same on both large and small animals. All the marks noted on the proximal tibia were of one type (Binford 1981:118, Tp-2); that is, marks along the edge of the medial tuberosity parallel to the proximal articular surface. Similarly, all the marks noted on the distal ends of the tibia were across the tip of the medial malleolus, as shown in Binford (1981:118, Td-3). No marks related to filleting or the removal of meat were noted on either the femur or the tibia.

SCAPULA

In many ways the scapula appears analogous to the pelvis in that of the parts of the upper-front leg it exhibits a biased high frequency of cut or dismemberment marks on bones from the smaller-body-size animals (see

TABLE 4.17

Proximal Tibia: Cut, Hack, and Gnaw Marks

		Cut marked		*Hack marked*		*Gnawed*	
Bovid class	*MNE* (1)	*No.* (2)	*%* (3)	*No.* (4)	*%* (5)	*No.* (6)	*%* (7)
I	13	2	.15	0	0	0	0
II	10	1	.08	0	0	0	0
III	19	6	.32	0	0	1	.05
IV	7	1	.14	0	0	2	.29
V	2	0	0	0	0	1	.50
	51	10	.20	0	0	4	.08

TABLE 4.18

Distal Tibia: Cut, Hack, and Gnaw Marks

		Cut marked		*Hack marked*		*Gnawed*	
Bovid class	*MNE* (1)	*No.* (2)	*%* (3)	*No.* (4)	*%* (5)	*No.* (6)	*%* (7)
I	22	3	.14	0	0	0	0
II	21	6	.29	0	0	0	0
III	45	1	.02	0	0	0	0
IV	41	8	.20	0	0	4	.10
V	6	1	.17	0	0	0	0
	135	19	.14	0	0	4	.03

Table 4.19). Unlike the pelvis, however, there is a relatively high frequency of hack marks on the scapulae of large-body-size animals.

In terms of the types of cut marks, there is a total contrast between the marks on the bones from the small body-sizes (classes I and II) versus the larger classes. Four types of cut marks were observed on the scapulae of the small animals.

1. Thirteen of the marked pieces showed transverse marks just below the lip of the glenoid fossa on the dorsal surface (see Binford 1981:122, mark S-1).
2. Eight marks were noted on the ventral surface in an analogous position; that is, just below the lip of the glenoid fossa. These would be inflicted after the front quarter was removed from the animal and the scapula was being disarticulated from the humerus.

TABLE 4.19

Scapula: Cut, Hack, and Gnaw Marks

		Cut marked		*Hack marked*		*Gnawed*	
Bovid class	*MNE* (1)	*No.* (2)	*%* (3)	*No.* (4)	*%* (5)	*No.* (6)	*%* (7)
I	103	23	.22	0	0	0	0
II	50	7	.14	0	0	0	0
III	61	5	.08	3	.05	6	.10
IV	35	1	.03	2	.06	9	.26
V	7	1	.14	2	.29	4	.57
	256	37	.14	7	.03	19	.07

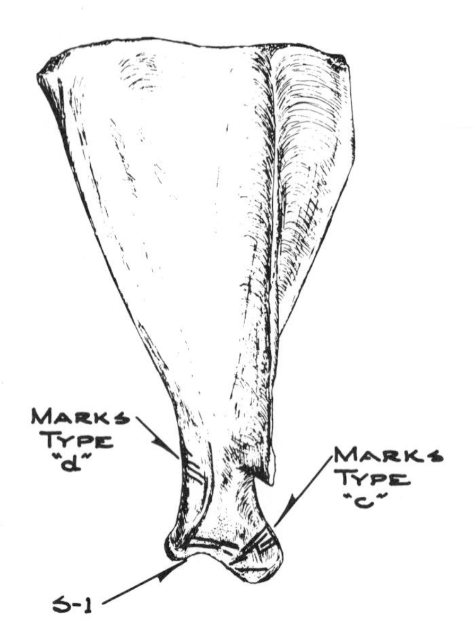

Figure 4.14 Dismemberment marks on the scapula.

3. Six had rather deep parallel marks diagonally across the supraglenoid tubercle, running from the scapula notch to the upper edge of the glenoid fossa (see Figure 4.14). This was a position not previously in my studies of cut marks. It is certainly produced when the scapula–proximal-humerus joint is heavily flexed, and the cut is inflicted while cutting the supraspinatus muscle off the head of the humerus. The animal would have to be quite flexible when such a cut was inflicted (see Figure 4.14).
4. Three scapulae had small, short, chevron marks on the upper axillary border, roughly opposite the scapula notch (Binford 1981:122; S-2) under the metacromion of the scapula spine. These marks would be inflicted either when the joint was fully extended (not flexed) or during the removal of meat from the scapula (e.g., filleting).

The contrast to the larger-body-size animal could hardly be greater. First, all the marks on bones of the larger animals that were inflicted during dismemberment were hack marks, except one that was like Type 1 above—mark S-1. The remaining six marks noted on large-animal scapulae were filleting marks, two of which were longitudinal on the ventral surface of scapula blades (see Figure 4.15), and the remainder were longitudinal marks along the infraspinous fossae (Binford 1981:98, S-3). I think it is fairly certain that these filleting marks were inflicted at the site after meat-covered scapulae had been introduced. This means that all the dismemberment

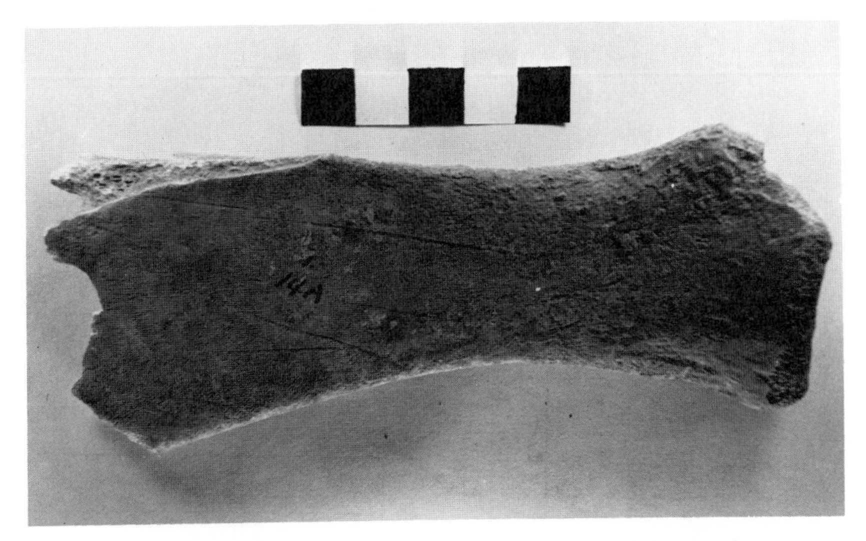

Figure 4.15 Filleting marks on the ventral surface of the scapula.

marks observed on the large-animal scapulae (except one) were inflicted by choppers, whereas all those on the smaller animals were inflicted by cutting–slicing tools. The dominance of chopping marks corresponds, as repeatedly noted, to the high frequency of animal-gnawed pieces. This continues to be true for the scapula.

HUMERUS AND RADIOCUBITUS

Table 4.20 summarizes the information regarding cut, hack, and graw marks observed on the articular ends of the upper-front limb bones. An interesting pattern is clear. Only one example of hacking or chopping was observed on all the other bones from the upper-front limb. This is in marked contrast to the situation with the scapula, but is similar to the situation with all the bones of the upper-rear leg in which no examples of hacking or chopping were noted on the proximal femur down through the distal tibia. Many more cut marks remaining from dismemberment were noted on the pelvis and proximal femur of small animals, and a parallel situation is seen with the scapula.

Unlike the situation with the rear leg, where little difference is seen in the numbers of dismemberment marks on the articular ends of the upper-limb bones, there is a concentration of dismemberment marks on the radius of the smaller animals, suggesting that there was a regular disarticulation of the smaller animals at the distal humerus–proximal radiocubitus joint, whereas no such regular dismemberment at this joint appears to character-

TABLE 4.20

Inflicted Marks: Upper-Front Limbs

		Cut marked		*Hack marked*		*Gnawed*	
Bovid class	*MNE* (1)	*No.* (2)	*%* (3)	*No.* (4)	*%* (5)	*No.* (6)	*%* (7)
		PROXIMAL HUMERUS					
I	6	0	0	0	0	0	0
II	4	0	0	0	0	1	.25
III	4	0	0	0	0	0	0
IV	7	2	.29	0	0	3	.43
V	2	0	0	0	0	1	.50
	23	2	.09	0	0	5	.22
		DISTAL HUMERUS					
I	37	6	.16	0	0	0	0
II	40	2	.06	0	0	0	0
III	26	0	0	0	0	0	0
IV	27	7	.27	0	0	0	0
V	9	4	.44	0	0	0	0
	139	19	.14	0	0	0	0
		RADIUS					
I	12	3	.25	0	0	0	0
II	14	3	.21	0	0	3	.21
III	12	4	.33	0	0	0	0
IV	24	0	0	1	.04	0	0
V	11	0	0	0	0	1	.09
	73	10	.14	1	.01	4	.05
		CUBITUS					
I	14	2	.14	0	0	0	0
II	8	0	0	0	0	2	.25
III	8	0	0	0	0	2	.25
IV	21	2	.10	0	0	1	.05
V	8	1	.13	0	0	1	.13
	59	5	.08	0	0	6	.10
		DISTAL RADIOCUBITUS					
I	2	0	0	0	0	0	0
II	8	0	0	0	0	0	0
III	15	2	.13	0	0	0	0
IV	20	1	.05	0	0	0	0
V	2	0	0	0	0	0	0
	47	3	.06	0	0	0	0

ize the larger animals. This pattern complements a shift in the relative frequencies of upper- versus lower-limb parts represented from the small- versus large-body-size classes (see Table 3.6).

Gnawing shows a very interesting pattern. It is present with increasing frequency on both the scapula and proximal humerus of larger animals. On the other hand, the distal humerus and proximal radius show little animal gnawing, although there is some gnawing on the olecranon. Gnawing is absent on the distal radiocubitus. All in all, gnawing is not a common characteristic except on the scapula and proximal humerus. Bones down the leg from this area of the front limb appear untouched by animals. This is in contrast to the rear leg, where gnawing was a consistent characteristic on all large-animal upper-limb bones, except the distal tibia. There is very clearly a biased gnawing on the upper-rear leg.

It should be pointed out that this is the area of a freshly killed animal that almost always receives the initial attention from feeding animals.

> A lion usually eats the hindquarters first, followed by the forequarters and lastly the head. . . . leopards, too often eat viscera first, but cheetah reject them and tigers tend to consume the meat from the rump and thighs before the intestines. (Schaller 1972b:269; © 1972 by the University of Chicago)
>
> When the kill was a domestic cow or buffalo, the tiger began to eat at the rump in all of the instances observed, although it sometimes started at the neck if another tiger had already occupied the preferred place at the hindquarters. (Schaller 1967:297)
>
> If the carcass is intact, however, or if the victim is still alive after being hunted down, they [spotted hyaena] tear open the belly and loins; the first parts to go are usually the testicles or udder. . . . Once the abdomen is opened, the entrails are pulled out and the soft parts eaten. . . . At this stage the hyenas begin eating the abdominal and leg muscles and the skin. . . . Once a large part of the muscles has been consumed, the hyenas are able to start tearing off legs, and one sees the first animals running off with their loot. (Kruuk 1972:126; © 1972 by the University of Chicago)

(See also my summary of analogous documents in Binford 1983c.)

Tool-inflicted cut marks also exhibit an interesting distribution. As already noted, the marks on the scapulae from the larger animals were inflicted during filleting, while dismemberment was accomplished by hacking. Dismemberment marks dominate those on the humerus. The two marks noted on the proximal humerus were across the greater trochanter (see Binford 1981:123, Figure 4.30c, mark Hp-2). Marks on the distal humerus are all marks familiar from previous studies. Eight marks were multiple, parallel cuts on the margins of the radial fossa (posterior face), and seven more were transverse marks across the posterior face, just above the radial fossa. Three more of the marks were transverse across the face of the distal condyle. (All these are illustrated in Binford 1981:113, Figure 4.30.) These are all marks made on the posterior face, just above the plane of the radius.

They are believed to represent the same cut, but with the joint flexed in different positions. The transverse cut across the posterior face above the radial fossa (Binford 1981:123, Figure 4.30e, mark Hd-2) is believed to have been inflicted when the joint was radically flexed, as is common on stiff carcasses. This mark is restricted to bones of *Taurotragus* (eland) represented in the deposit (size class IV). Of course, this is an animal that has consistently yielded evidence of having been butchered when stiff.

Two additional cuts were noted in the distal humerus from moderate-body-size animals (size classes III and IV). One was the angular mark across the face of the medial epicondyle (shown in Binford 1981:123, Figure 4.30f). This is the same mark illustrated by H. Bunn (1981) on a bone from Koobi Fora. This mark is believed to be most commonly produced when the joint is flexible and fresh. No marks of this type were noted among bones from the large body-size classes. It should be pointed out that all the marks on the humerus were dismemberment marks.

Bones from the small animals (size classes I and II) yielded a similar pattern. Two marks were noted on the lateral face of the cubitus, diagonally placed with respect to the semilunar notch (see Binford 1981:125, Figure 4.32a, mark Rcp-2). A related dismemberment mark was observed on the proximal radius from animals in size classes I and II (Binford 1981:125, Figure 4.32b, mark Rcp-5). This mark is a consequence of the same dismemberment tactics as those summarized for the distal humerus.

In marked contrast were the marks recorded for animals of the larger body sizes. Short chevron marks (see Binford 1981:126–134) are generally inflicted during filleting operations. Four of these were observed on the proximal ends of radii from animals in size class III, and one such mark was observed on a *Tauratragus* radius (size class IV), whereas no dismemberment marks were observed on the proximal radius from larger animals. One small slice mark was noted on the apex, or end, of the olecranon but no marks were typically noted around the semilunar notch where dismemberment is commonly indicated. This pattern was continued for the distal end of the radius from large animals, in which one encircling cut was observed some 5 cm (2 inches) up the shaft on the medial face, with a similar mark on the posterior face of the shaft at the distal end. In addition, multiple short chevrons were observed on the anterior face of the distal radius (see Binford 1981:133, Figure 4.39, mark Rcp-3).

The contrast could not be greater: all the marks on bones of small-body-size animals were dismemberment marks, whereas all the marks on the bones of the larger animals most likely were inflicted during filleting or the removal of tissue from the bones. This difference is further emphasized when it is noted that all the radius bones from the larger-size animals were regularly impacted with percussion blows on the anterior face just below the proximal end of the radius, whereas only 4 of the 26 bones from the animals in body-size classes I and II are broken. In the latter case, it is common for there to be a break through the shaft approximately 5 cm (2 inches) above the distal end of the radius. It is likely that this break was made during dismemberment rather than marrow recovery, whereas marrow recovery was clearly the context for breakage for the bones of the larger animals. The dismemberment interpretation is further supported by the strong reduction in numbers of distal articular ends from small-body-size animals—which were apparently left in the field where the lower limbs were abandoned.

The conclusion to be drawn is that small animals were commonly dismembered by cutting through the proximal radius–distal humerus joint or by breaking through the shaft of the radius. Correlated with this pattern is a general absence of lower legs in the Klasies site. Presumably they were regularly abandoned in the field. In marked contrast are the bones from the larger animals, in which dismemberment by cutting through the articulation between the distal humerus–proximal radiocubitus is sometimes indicated. The major marks remaining on the radiocubitus were inflicted during the removal of tissue and the subsequent breakage for marrow. The meat on the radiocubitus is minimal, and it seems most likely that the removal of tissue was related to the preparation of the bone for marrow removal rather than filleting of the upper limb, particularly since no filleting marks were noted on the humerus.

The pattern of gnawing might be taken as suggestive that with the large animals, the front limb was commonly recovered from kill–death locations in two separate parts. The scapula with attached proximal humerus was commonly dry and stiff, particularly on exposed areas. Chopping and hacking on exposed surfaces may have been followed by some filleting of the meat from the scapula itself. The other unit would be the distal humerus with attached lower-leg parts, including the phalanges. Gnawing and exposure were localized around breaks through the humeral shaft, whereas skin and tissue of the lower-leg parts would have been largely untouched by carnivores. The distal humerus would have been removed by cutting through the distal humerus–proximal radiocubitus articulation, and then the bones of the lower leg successively prepared for marrow extraction.

A similar picture seems to apply to the rear leg, only the most common segment recovered and transported to the site was the lower-rear leg with attached distal tibia. Breakage through the tibial shaft would have been accomplished prior to the introduction of the part to the site, perhaps by nonhominid predator scavengers. Such a view is supported by the high proportion of animal-gnawed parts of the upper-rear leg above the distal tibia, parts that are rather infrequent at the site. This upper- versus lower-rear-limb contrast is analogous to the distal-scapula-and-proximal-humerus versus the lower-leg comparison noted for the front leg. A marked contrast is seen in the anatomical part frequencies of ungnawed upper-leg meat-yielding parts, whereas large animals are represented by low frequencies of gnawed upper-leg parts and high frequencies of lower-leg parts, all heavily processed for marrow.

The Lower Limbs

The most striking single feature of the lower limbs is the pattern of bone breakage characteristic of the bones from the larger animals. Thus the descriptive task is increased considerably in this section, because breakage must be treated more completely, along with cut, chop, and gnawing marks. The reader is already familiar with the description of inflicted marks, so I have chosen to describe them first and then to proceed to discussion of breakage observed among metapodials and phalanges.

CARPALS AND METACARPALS

The very small carpal bones are apt to have passed through the ½-inch screens that were reportedly used for sieving the deposits at Klasies River Mouth. Not surprisingly no carpals are referred to body-size class I (small) and only four (not listed in table) are identified to body-size class II. I think

TABLE 4.21

Carpals: Frequency and Modification, Tabulated by Body Size

	Bovid class									
	I		*II*		*III*		*IV*		*V*	
Carpals	*MNE (1)*	*MKD*[a] *(2)*	*MNE (3)*	*MKD (4)*	*MNE (5)*	*MKD (6)*	*MNE (7)*	*MKD (8)*	*MNE (9)*	*MKD (10)*
Scaphoid	0	0	0	0	13	0	30	0	4	0
Lunate	0	0	0	0	11	0	26	3	3	0
Cuniform (manus)	0	0	0	0	14	1	31	5	6	1
Magnum plus trapezoid	0	0	0	0	10	1	22	2	4	1
Uniceform	0	0	0	0	10	4	25	5	4	3
Σ					58	6	159	15	21	5
MAU					5.8		15.9		2.1	

[a] MKD, Marked (all modifications).

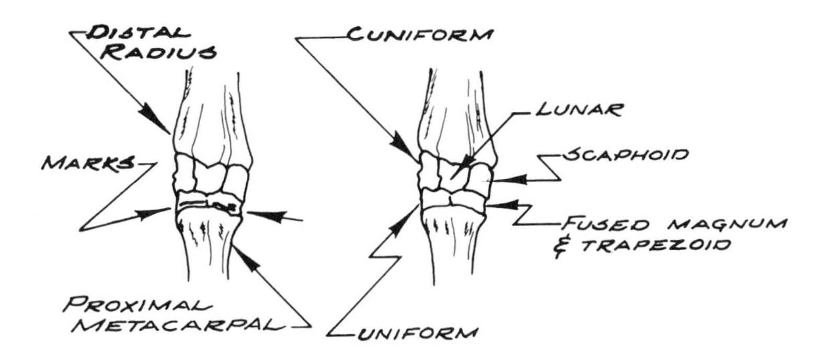

Figure 4.16 Cut marks on carpals (right anterior view).

it is fair to dismiss these body parts as biased by the recovery techniques. Table 4.21 summarizes the frequencies and numbers of modifications noted on carpals, tabulated by body-size class.

No chop marks and no gnawed carpal bones were observed. Slicing marks were present but these tended to be concentrated across (transverse marks) the anterior face of the cuniform, and the fused mangum and trapezoid (Figure 4.16). These marks roughly parallel the anterior edge of the metacarpal and are marks most commonly made when a pressure flex can be accomplished at this joint. Once the joint is forced into a flexed position, a short cut across the lower anterior face of the bent joint pops the articulation with the proximal metacarpal loose so that the joint may be twisted and the tissue at the rear can be cut without impacting bone. This procedure results in transverse cut marks across the posterior face of the metacarpal, just below the lip of the articular surface.

In my experience, this type of disjointing is almost exclusively associated with dismemberment when the upper limb has already been removed from the body and may be rested on the ground. The metacarpal is then bent down against the radiocubitus, pressure flexing the articulation of the carpus, making a cut under such tension quite effective.

On the other hand, when the lower limb is removed while the upper limbs are still attached to the body, it is more common for a transverse cut to be made along the medial face just below the rim of the proximal metacarpal. The joint is then twisted out toward the lateral face and a second transverse cut made along the crest of the lateral posterior edge of the articular rim.

The cut marks on the proximal metacarpal (Table 4.22) from the two smallest body-size classes (I and II) are all dismemberment marks, and are

TABLE 4.22

Proximal Metacarpals: Cut, Hack, and Gnaw Marks

Bovid class	MNE (1)	Cut marked No. (2)	Cut marked % (3)	Hack marked No. (4)	Hack marked % (5)	Gnawed No. (6)	Gnawed % (7)
I	5	1	.20	0	0	0	0
II	11	2	.18	0	0	0	0
III	20	2	.10	0	0	0	0
IV	50.7	16	.32	1	.02	4	.08
V	9	2	.27	3	.33	2	.22
	95.7	30	.31	4	.04	6	.06

all combinations of transverse cuts along the medial face just below the rim of the articular surface (three cuts), and the remaining four are transverse marks across the posterior surface just below the edge of the articular surface. No marks were observed on the distal metacarpals from the smaller animals (Table 4.23).

Bones referable to the three largest body-size classes (III–V) contrast with the smaller in a number of ways. First there were hack or chop marks on the proximal ends. Similarly, there was regular animal gnawing on the large-animal bones. Finally, while all the marks noted on the small-animal bones were dismemberment-related, there were substantial numbers of short chevron marks on large-animal metacarpals that almost certainly were inflicted during the removal of skin and tissue. These contrasts reflect differences in the initial conditions and treatment of the larger animals versus the smaller—a contrast maintained in the details of the cut marks. For instance, the cut marks very close to the joint between the metacarpal and the small bones of the carpus are infrequently among the larger animals. Typically, there is a transverse cut mark about 2 cm (1 inch) below the lip of the proximal articular surface across the anterior face (three marks observed). This condition is commonly paired with a mark across the posterior face, also about 2.5 cm below the rim of the articular surface (seven marks observed). The same placement was noted in all the hacked or chopped examples.

It is believed that these marks on the shaft below the articulation between the metacarpal and the carpals is directed toward skinning the joint preparatory to disjointing, primarily with the use of leverage. If an animal is stiff and the skin is relatively dry, it acts as a binding sheath, which must be removed before the stiff and rigid articulation may be cracked apart. Only one of the marks observed on the larger animals was a cut mark across the

TABLE 4.23

Distal Metacarpals: Cut, Hack, and Gnaw Marks

		Cut marked		*Hack marked*		*Gnawed*	
Bovid class	*MNE (1)*	*No. (2)*	*% (3)*	*No. (4)*	*% (5)*	*No. (6)*	*% (7)*
I	4	0	0	0	0	0	0
II	10	0	0	0	0	1	.10
III	17	0	0	0	0	0	0
IV	50	9	.18	1	.02	4	.08
V	8	1	.13	0	0	0	0
	89	10	.11			5	.06

medial face close to the rim of the articulation, which, as was pointed out, derived from dismemberment when the lower limbs were supple and still attached to the body. Similarly, only two marks were transverse and close to the rim across the posterior face, the typical position of cuts inflicted using a tension flexed joint prior to cutting.

That the larger animals had the tissue removed and were regularly skinned is shown by no less than six examples of short chevron marks on the anterior face of the metacarpal shaft and another three examples on the lateral face at least 3.8 cm (1.5 inches) below the articular end. Such marks are well documented as produced during the removal of tissue from the bones. Because there is essentially nothing on a metacarpal except skin and bundles of tendon in the posterior groove, the short chevrons were most certainly inflicted during skinning operations.

Skinning of freshly killed animals is generally accomplished by making an encircling cut around the distal metapodial (see Binford 1981:107) and then a quick incision down the leg on the inside to the crotch, followed by stripping the skin down the leg (see Figures 4.17 and 4.18). When an animal is flexible and supple, skinning is comparatively easy and need not be accomplished by short slicing cuts under the skin between the underlying tissue and the fascia. Such cuts are, however, common when the skin has dried out—even if the skin has been soaked in water the original tendency to separate at the fascia is no longer maintained—and when the bone is being cleaned of attached tissue in preparation to controlled cracking for the removal of marrow. (See Binford 1978:152–156; 1981:150–163 for discussions of cleaning bones for marrow cracking.)

As has been pointed out with so many other properties, the marks and the placement of marks on the metacarpals is consistent with the processing for marrow of dried and relatively stiff lower limbs from large animals.

Figure 4.17 Cutting the skin down the inside of a sheep leg.

Tarsals and Metatarsals

As was the case with the carpals and metacarpals, the clues to the dismemberment strategies are of prime importance. There are three basic approaches to the dismemberment of the complex joints and articulations between the distal tibia and the proximal metatarsal. When an animal is fresh and supple, the easy disjointing strategy (Strategy 1) is disarticulation between the distal tibia and the astragalus. In this situation there are essentially three cuts that are made, coupled with the use of leverage to accomplish this disjointing.

Figure 4.18 Caribou, showing skin ripped down one leg during skinning procedure.

STRATEGY 1

1. The first cut is across the anterior face of the astragulus, in the angle of the bend of the leg (Figure 4.19). This cuts through a bundle of tendons, including the tibialis anterior, as well as a number of extensor tendons ultimately attached to the phalanges. This cut is frequently the first one made because it is easy to place given the normal small flexed angle at which the leg is positioned when at rest. This commonly produces cut marks across the anterior face of the astragulus (Binford 1981:120, Figure 4.27e, mark TA-1). The same cut may impact the calcaneus on the medial ridge that articulates with the lateral face of the astragalus (Binford 1981:120, Figure 4.27a and b, mark TC-1). This cut may be made regardless of whether the leg is supple or relatively stiff.

2. The next step is to sever the calcaneal or achilles tendon, which is attached to the tuber calcis or the extreme posterior end of the calcaneus. This act may result in short, nicklike cuts on the distal end of the calcaneus, and sometimes marks across the dorsal ridge of the body of the calcaneus as the knife impacts bone after having cut the tendons (Binford 1981:120, Figure 4.27b and c, mark TC-3).

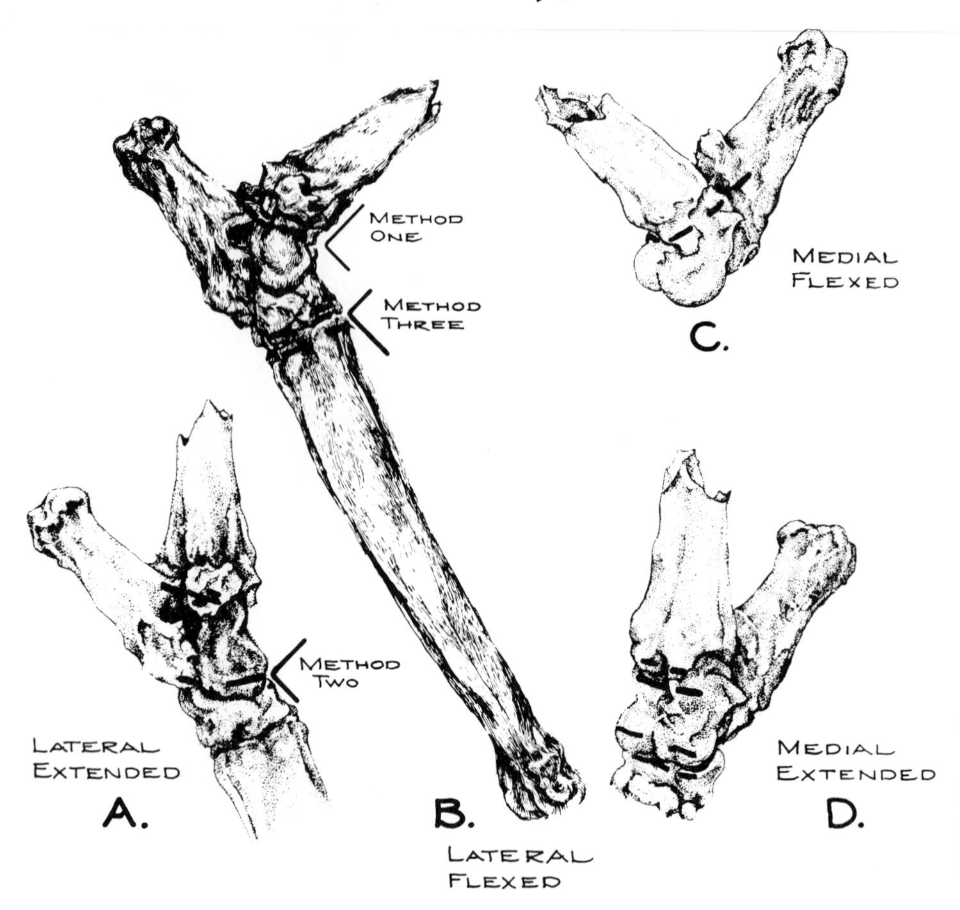

Figure 4.19 Alternative disjointing strategies for the left leg when the joint is either stiff or flexible—metatarsal, tarsal, and distal tibia articulations.

3. If the carcass is supple and flexible, the joint can now be manipulated into the position shown in Figure 4.19B. In the flexed position, cuts may be made across the distal tip of the tibia (the interior or medial malleolus). After this cut, the same type is made on the exterior or lateral face of the joint, where the knife may scar the lateral malleolus (this is an unfused remnant end of the fibula) and may also nick the margins of the proximal edge of the astralagus. The joint may now be levered apart, resulting in the lateral malleolus remaining with the distal tibia while all tarsals remain articulated and attached to the proximal metatarsal. With this type of disarticulation, marks occur on the lateral and medial malleolus, with the possibility of some nicking on the medial face of the astragalus orientated in a longitudinal or only slightly diagonal plane. These marks, of course, should covary with nick marks on the tuber calcis and with transverse marks across the anterior face of the astragalus.

When the joint is stiff and more difficult to manipulate, there are two other disjointing strategies:

STRATEGY 2

The second strategy may be followed when it is deemed desirable not to cut the calcanial tendons and to leave at least the calcaneus and astragalus attached to the distal tibia (Figure 4.19). Among modern hunters this is commonly done if the meat is to be dried and the ham hung up as a unit. In that case the calcanial tendons are used as a hanging hook.

The first cuts are generally the same as in Strategy 1—namely, a transverse cut across the face of the anterior astragalus, which may be extended around the joint, catching the rim of the calcaneus on one side and the proximal medial tip of the navicular cuboid on the medial face. Leverage is then used to pull the joint apart at the juncture of the astragalus–calcaneus with the navicular cuboid. The latter then remains attached to the proximal metatarsal. The parts remaining attached to the tibia are as shown in Figure 4.19C.

STRATEGY 3

The third strategy is followed most commonly when the joint is totally inflexible, and/or when the metatarsal is being specifically disarticulated for purposes of marrow cracking. A cut is made, essentially circling the medial–anterior–lateral face of the joint at the plane of articulation between the proximal metatarsal with the navicular–cuboid and the ectocuniform. This results in transverse marks across the face of the medial–anterior and lateral margins of the proximal metatarsal, as well as similar transverse marks across the navicular cuboid and the ectocuniform. Actual disjointing may be

TABLE 4.24

Cut Marks on Tarsals, Distal Tibiae, and Proximal Metatarsals[a]

	Bovid class									
	I		*II*		*III*		*IV*		*V*	
	MNE	*MKD*	*MNE*	*MKD*	*MNE*	*MKD*	*MNE*	*MKD*	*MNE*	*MKD*
Exterior malleolus	0	0	0	0	39	2	16	2	6	1
Calcaneus	25	0	20	0	34	1	31	3	21	1
Astragalus	17	3	21	7	42	13	68	12	21	5
Navicular–cuniform	5	2	3	1	17	7	65	21	8	1
Ectocuniform	0	0	0	0	38	6	70	7	9	0
	MAU	% *MKD*	*MAU*	% *MKD*	*MAU*	% *MKD*	*MAU*	% *MKD*	*MAU*	% *MKD*
Interior malleolus	11.0	.24	10.5	.29	22.5	.02	20.5	.20	3.0	.17
Exterior malleolus	0	0	0	0	19.5	.05	8.0	.13	3.0	.17
Calcaneus	12.5	0	10	0	17.0	.03	15.5	.10	10.5	.05
Astragalus	8.5	.18	10.5	.33	21.0	.31	34.0	.18	10.5	.24
Navicular–cuniform	2.5	.40	1.5	.33	8.5	.41	32.5	.32	4.0	.13
Ectocuniform	0	0	0	0	8.0	.16	35.0	.10	4.5	.22
Proximal Metacarpal	3.5	0	10.5	0	7.5	.40	19.5	.26	7.0	.21

[a] MKD, marked. Note: dashed line indicates inferred placement of dismemberment cuts.

accomplished by leverage followed by a transverse cut across the posterior face just below the articular surface articulation.

Table 4.24 summarizes the data from the tarsals, together with data on cut mark frequencies from the distal tibia, as well as the proximal metatarsal. Thus the relative roles of the above three strategies may be assessed.

Examination of the frequency information illustrates that among the small animals the cuts were most commonly made across the distal face of the astragalus also impacting the navicular–cuboid (Strategy 2). This positioning also corresponds to the major drop in frequency between adjacent parts that, other things being equal, almost always indicates the point of dismemberment. Clearly within size classes I and II dismemberment was by Strategy 2. This results in the calcaneus and astragalus as well as the lateral malleolus being removed in articulation with the distal tibia.

In the case of the small-animal classes, the metatarsal together with phalanges and attached navicular cuboid were discarded prior to being returned to the site. This method of butchering is most common when the ham is to be hung and when there is an attempt to prevent flies or extraneous matter from invading the meaty areas. With this type of butchering, the skin remains firmly gripped around the tarsals and the meat of the tibia and the lower femur is covered by skin.

I have not observed this type of butchering associated with the use of carrying devices such as that shown in Figure 4.20. Use of poles on which the ham is suspended by the calcanial tendons is a possibility, however; animals of size classes I and II would be small enough to be carried whole from kill to camp, much in the manner shown in Figure 4.21. (See also an excellent photograph [in Howell 1965:187] of a modern Bushman carrying a steenbok obviously butchered in the manner suggested here.) The butchering tactics indicated are more likely related to cooking tactics.

Among hunters cooking is commonly accomplished in an ash oven:

> meat is baked under hot ashes next to the fire as this does not dry out the meat. When the strips of meat are cooked, they are removed and the ashes and sand beaten off them with a stick. . . . even the heads of most animals are also baked and eaten. . . . the meat of smaller animals is baked in special ways. Porcupines, for instance, are covered with grass which is burnt to singe the hair. . . . After this the porcupine is baked whole in the hot ashes of a large fire which is made in a shallow trench to one side of the huts. . . . As a rule all animals are skinned and cut open and their intestines removed. An exception to this rule is made when the young duiker or steenbok are cooked. These animals are cooked whole. They are not skinned, nor are the intestines removed, as young kids live on milk only and the contents of their intestines are not considered . . . to be undesirable. . . . Tortoises are killed with a knife and baked on their backs in the hot ashes. (Steyn 1971:284)

Cooking in the skin is not done so much because hunting and gathering peoples are averse to eating food covered with ash. In fact, many foods are

Figure 4.20 !Kung bushmen carrying processed *biltong* from a kill to a residential camp.

cooked directly in the ash as suggested in the above quote and as is illustrated in Figure 4.22. Cooking in the skin preserves nutrient juices and fats that otherwise tend to result in the roasting pit flaming up and the meat drying out, with the attendant loss of the fats and oils.

This principle was well described many years ago by Charles Darwin in *The Voyage of the Beagle* in describing cooking of meat by the gauchos of the Falkland Islands.

> He cut off pieces of flesh with the skin on it, but without the bones, sufficient for our expedition. We then rode to our sleeping-place, and had for supper "carne con cuero," or meat roasted with the skin on it. This is as superior to common beef as venison is to mutton. A large circular piece is taken from the back roasted on the embers with the hide downwards and in the form of a saucer, so that none of the gravy is lost. If any worthy alderman had supped with us that evening, "carne con cuero" without doubt would soon have been celebrated in London. (1839: 190–191)

This same lauded result of cooking meat in its skin is cited by Stow, paraphrasing early African explorers to the effect that the feet of large animals such as elephants or hippopotamus were cooked in ash ovens in their skins, yielding "a dish fit for an emperor" (Stow 1905:60).

Figure 4.21 Nharo Bushman carrying a steenbok (*Raphicerus*) back to camp.

It is quite likely that the methods of butchery indicated for the small animals were related to cooking procedures in which the upper leg was cooked in its skin. This would often result in the proximal femur's becoming singed and burned in the process, and as we will see later, in the section on burning, this is exactly the case.

The procedure of disjointing animals of size class III is more akin to that indicated for the small animals, with the exception that more metatarsals are recovered and they show butchering marks.

Tarsals from animals of size classes IV and V illustrate a pattern that is different from the small animals in a number of important ways. Strategy 1 (Figure 4.19) is indicated whenever the astragalus is removed with the metatarsal rather than with the tibia. This is particularly evident in the large-body-size animals, in which not only the astragalus, but also the calcaneus is attached to the metatarsal after dismemberment at the tarsals. The other important difference is that there are many more metapodials and fewer

Figure 4.22 Springhares being placed in an ash hearth in preparation for cooking by a Masarwa Bushman.

upper-limb parts at the site. This certainly results from the selective introduction to the site of lower-limb parts with attached tarsals, while the upper meat-yielding limb bones were abandoned before arrival on the site. In short, there is a shift in bias between the small and the large animals. With the former, the metatarsal is abandoned before introduction or, if introduced, is not processed for marrow, whereas with the large animals, the meat-yielding parts are missing at the site and a biased introduction of marrow-yielding bones dominates the frequencies of rear-leg parts. As illustrated, the shift in bias is associated with different dismemberment strategies. As further documentation of this pattern, Table 4.25 summarizes the information on cut marks together with hack and graw marks for the three larger tarsal bones. Unlike many of the earlier comparisons, there is a general absence of hack marks, and only on the posterior end of the calcaneus is there any measurable animal gnawing. The latter is generally restricted to the two larger body-size classes. The actual placement of cut marks on the astragalus and calcaneus are summarized in Table 4.26.

As shown in Figure 4.19, when the cut marks are low on the astragalus on either the medial or anterior face, the joint was most likely flexed when

TABLE 4.25

Cut, Hack, and Gnaw Marks on the Larger Tarsal Bones

Bovid class	*MNE* (1)	*Cut marked* No. (2)	*Cut marked* % (3)	*Hack marked* No. (4)	*Hack marked* % (5)	*Gnawed* No. (6)	*Gnawed* % (7)
		NAVICULAR-CUBOID					
I	5	2	.40	0	0	0	0
II	3	1	.33	0	0	0	0
III	17	7	.41	0	0	0	0
IV	35	11	.31	3	.09	0	0
V	8	1	.13	0	0	0	0
	68	21	.31	3	.04	0	0
		ASTRAGALUS					
I	17	3	.18	0	0	0	0
II	21	10	.33	0	0	0	0
III	42	13	.31	0	0	0	0
IV	68	12	.18	0	0	3	.04
V	21	5	.24	0	0	0	0
	169	40	.24	0	0	3	.02
		CALCANEUS					
I	25	0	0	0	0	0	0
II	20	0	0	0	0	0	0
III	34	1	.03	0	0	0	0
IV	31	3	.10	0	0	3	.10
V	21	1	.05	0	0	1	.05
	131	5	.04	0	0	4	.03

the cut was made, whereas a semiflexed or straight position is indicated when the cuts are across the middle of the astragalus. It is very clear that small animals were more commonly butchered with the joint flexed, whereas the larger animals were dismembered with the joint in a "natural" position—that is, the position that the joint assumes naturally when it is allowed to dry and become stiff (see Figure 4.19). The regular shift to medial placement strongly indicates (1) removal of the metatarsal with attached tarsals, and (2) the joint's being relatively inflexible at the time of dismemberment.

METATARSALS

Cut marks were common on the proximal metatarsals and hack marks were consistently present on the metatarsals of the larger animals. Table 4.27 summarizes the data on modifications noted on metatarsal fragments.

TABLE 4.26

Cut Marks on the Astragalus and Calcaneus

	Astragalus						*Calcaneus*		
	Medial face			*Anterior face*					
Bovid class	*Total (1)*	*Middle (2)*	*Low (3)*	*Total (4)*	*Middle (5)*	*Low (6)*	*Dorsal ridge (7)*	*Lateral face (8)*	*Medial (9)*
I	3	0	3	0	0	0	0	0	0
II	10	3	7	2	0	2	0	0	0
III	11	9	2	5	3	2	1	0	0
IV	6	6	0	10	10	0	1	1	1
V	3	3	0	2	2	0	0	0	1

TABLE 4.27

Metatarsals: Cut, Hack, and Gnaw Marks

Bovid class	*MNE* (1)	*Cut marked* No. (2)	*Cut marked* % (3)	*Hack marked* No. (4)	*Hack marked* % (5)	*Gnawed* No. (6)	*Gnawed* % (7)
			PROXIMAL METATARSAL				
I	7	0	0	0	0	0	0
II	21	0	0	0	0	0	0
III	15	6	.40	1	.05	1	.05
IV	39	11	.28	7	.18	1	.03
V	14	3	.21	1	.07	3	.21
	96	19	.20	9	.09	5	.05
			DISTAL METATARSAL				
I	12	0	0	0	0	0	0
II	18	0	0	0	0	2	.11
III	14	2	.14	0	0	0	0
IV	26	3	.12	2	.08	1	.04
V	16	0	0	1	.06	0	0
	86	5	.06	3	.03	3	.03

Most of the cut marks on the proximal metatarsal were from dismemberment: 65% were transverse marks on the anterior face just below the parallel to the rim marking the transition to the articular surface. These marks are directly analogous to those described for the metacarpal. In size class III the 6 cut marks were of this type, in class IV, 11 were of this type, but in three cases there were also short chevrons indicative of filleting on the same bones marked during dismemberment. In the large animal class, 2 of the 3 marks noted were transverse and just below the articular rim. All other marks were short chevrons, generally occurring in pairs located down on the anterior face of the shaft, approximately 5 cm (2 inches) below the articular rim. Of all the hack marks, 7 were diagonal with respect to the long axis of the bone and placed between 1.25 and 5 cm (.5 and 2 inches) below the articular surface. This effect seems to be related to hacking through dried and tough skin, because given such a placement, disarticulation does not seem to have been the object.

Distal metatarsals had fewer cut-marked pieces than did the proximal parts, and all were from medium to medium–large bovids. In the case of the two marked pieces from size class III, both were transverse marks located up the shaft approximately 5 cm above the epiphysis and on the ventral or posterior face of the bone. These were probably inflicted during skinning.

Figure 4.23 Hack marks across the distal condyles of the metacarpal of an eland (*Taurotragus*).

One of the three marked pieces from size class IV was of this type, whereas the other two were marked across the lateral face of the distal condyle. The latter were certainly inflicted while dismembering the phalanges from the metatarsal.

Hack marks noted on the distal ends were all placed across the distal condyles and were certainly inflicted during dismemberment of the metatarsal–first phalange articulation. Figure 4.23 shows a heavy hack mark across the distal condyle of *Taurotragus*.

Gnawing was generally restricted to bones of animals from the larger body sizes. One interesting specimen of bushbuck (*Tragelaphus scriptus*) appears to have been gnawed by a hominid (Figure 4.24).

The presence of short chevron marks on the metatarsal as well as other

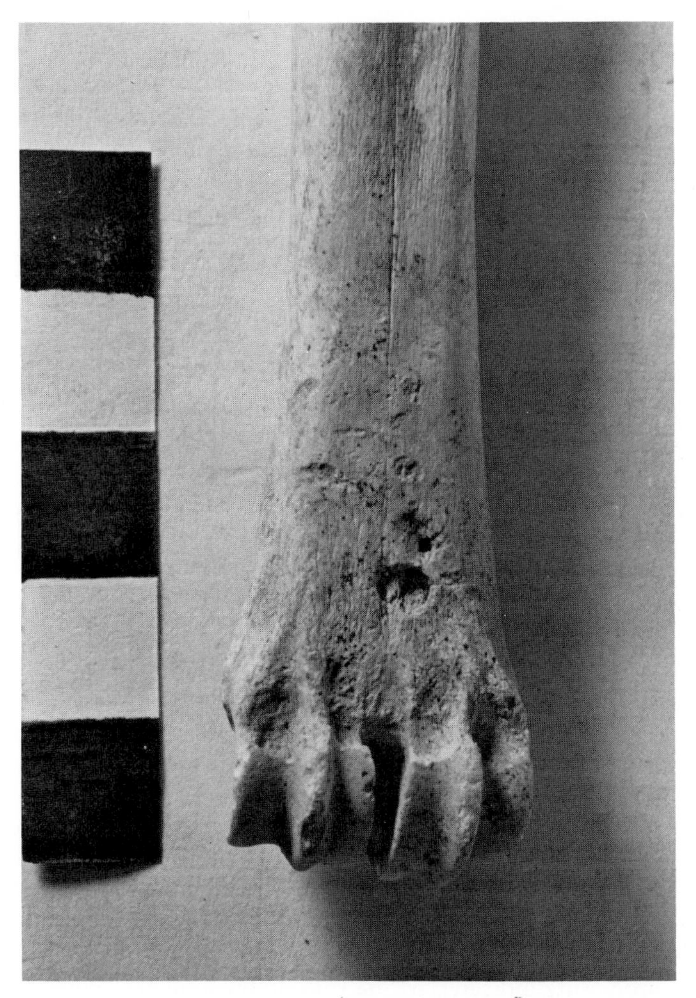

Figure 4.24 Distal metatarsal from a bushbuck (*Tragelaphus scriptus*), showing what are considered to be hominid tooth marks.

skinning marks illustrates that care was taken in the skinning of the metatarsals, which was also true for the metacarpals. These actions are most likely related to the preparation of the bone for marrow breaking. This interpretation is still further supported by the high frequency of dismemberment marks on the articular margins of the proximal metatarsal. Dismemberment at the juncture with the tarsals is a prerequisite to the easy breakage of the metatarsal through the proximal end, which is (as will be shown later in this chapter) the characteristic manner of marrow-bone breakage at Klasies River Mouth. The focus of dismemberment marks on the proximal

Figure 4.25 Tooth puncture on the first phalanx of an eland (*Taurotragus*).

metatarsal, as well as at the point of articulation of the distal tibia with the tarsals, most likely documents two separate acts of dismemberment—the former occurring in the field when the lower limb was removed from a carcass, and the latter cut made at the site of Klasies River Mouth when the metatarsal was being prepared for marrow cracking.

The third phalange was generally devoid of marks. Gnawing or tooth punctures (see Figure 4.25) were present on between 4 and 5% of the first and second phalanges of the larger animals.

The processing indicated by the cut and hack marks on the phalanges was almost certainly carried out in the site of Klasies, and is most certainly related to the processing of large-mammal lower limbs for marrow, including the phalanges themselves (as is shown in the following section on breakage).

Distinctive Breakage

Perhaps one of the most interesting, and certainly the most distinctive, pattern recognizable within the Klasies River Mouth fauna was the characteristic breakage of the metapodials and the phalanges from the medium to large animals. This pattern was not observed for any of the other long bones. Certainly if the breakage mechanics responsible for the distinctive pattern had been employed on the other bones, it would have been both identifiable and recognized. The conclusion is clear that a distinctive pattern of bone breakage was restricted to the processing of metapodials and phalanges from the larger-body-size animals.

METACARPALS

The metacarpal is generally the most robust of the metapodials, with thicker walls and a stronger proximal articular surface. In many bovids it is shorter than the metatarsal, and the marrow cavity is smaller relative to the total amount of bone present. Experiments with bones from domestic sheep as well as wild caribou have shown that, while the two species differ in the size bias favoring metatarsals, both display such a bias. In mature caribou the marrow cavity for metatarsals was 51 ml, whereas for metacarpals from the same animal it was only 21 ml (Binford 1978:24–25). It is my impression, however, that such extreme differences do not particularly characterize the species represented at Klasies River Mouth, although metatarsals are consistently larger. Table 4.28 summarizes the frequencies of recurrent fragment types observed among the metacarpals from four species representing the range of body sizes present in the total population. A number of extremely provocative patterns are illustrated in this table. First, for both metacarpals and metatarsals the number of unbroken articular ends is very high from the smaller-body-size animals. For instance, taking the complete bones, as well as bones only broken through the shaft with considerable segments of the shaft remaining (as would be the case from breakage during weathering), all metacarpal bones from *Raphicerus* fall into this class (Figure 4.26), whereas for *Taurotragus* there are no complete metacarpals and only 39% of all the proximal ends of metacarpals are represented by complete articular ends. The bones from the larger animal have been extensively processed and, as will be shown, broken to a distinctive pattern, whereas the bones from the smaller animals have not been generally processed for marrow.

In addition to this striking difference, there is a consistent difference in the fragmentation of the proximal and distal ends of the metapodials. The proximal ends from the large animals have been split and fragmented into segments. Although the metacarpal is generally split into roughly thirds (Figure 4.27), the more square shape of the proximal metatarsal has been split into halves, fourths, and even into small fragments representing essentially one-fourth to one-eights of the articular surface, when viewed from the proximal end (see Figure 4.28). On the other hand, the distal ends are

TABLE 4.28

Fragment Frequencies for Metapodials from Four Species of Different Body Size

	Raphicerus[a]			Tragelaphus[b]			Taurotragus[c]			Pelorovis[d]		
Metacarpals	*N*	*MNE*	%	*N*	*MNE*	%	*N*	*MNE*	%	*N*	*MNE*	%
Complete bones	3	3	.45–.60	0	0	0	0	0	0	0	0	0
Proximal end	4	4	.57	7	7	.78	23	23	.39	1	1	.11
Proximal ⅓	0	0	0	3	1	.11	110	36.7	.61	24	8	.89
Distal end	2	2	.40	9	9	1.00	52	52	.87	8	8	.89
Distal ½	0	0	0	0	0	0	0	0	0	0	0	0
Total MNE												
Proximal		7			8			59.7			9	
Distal		5			9			52.0			8	
Total MAU												
Proximal		3.5			4			29.85			4.5	
Distal		2.5			4.5			26.00			4.0	

Metatarsals												
Complete bones	4	4	.57–.50	2	2	.11	0	0	0	0	0	0
Proximal end												
Complete	3	3	.43	11	11	.52	9	9	.23	5	5	.36
Rear ½	0	0	0	5	5	.24	11	11	.29	4	4	.29
Rear ¼–⅛	0	0	0	0	0	0	35	11.5	.30	10	3.3	.24
Lateral ½	0	0	0	3	3	.14	8	8	.21	2	2	.14
Front ½	0	0	0	1	1	.05	19	19	.49	4	4	.29
Front ¼	0	0	0	0	0	0	5	2.5	.06	0	0	0
Distal end												
Complete	8	8	.67	6	6	.33	17	17	.65	6	6	.35
One-half	0	0	0	20	10	.63	18	9	.35	22	11	.65
Total MNE												
Proximal		7			22			38.5			14.0	
Distal		12			18			26.0			17.0	
Total MAU												
Proximal		3.5			11.0			19.5			7.0	
Distal		6.0			9.0			13.0			8.5	

[a] Grysbok: 9.07–13.6 kg (20–30 pounds).
[b] Bushbuck: 68 kg (150 pounds).
[c] Eland: 907 kg (2000 pounds).
[d] Giant buffalo: 1814 kg (4000 pounds).

Figure 4.26 Metacarpals of small bovids (mostly *Raphicerus*), showing lack of breakage.

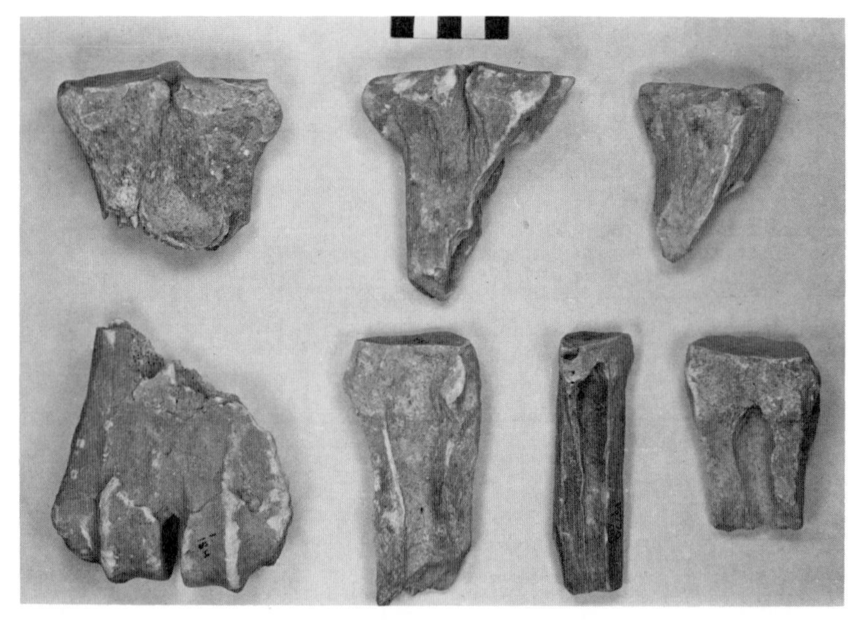

Figure 4.27 Metacarpals of giant buffalo (*Pelorovis*), showing distinctive breakage patterns.

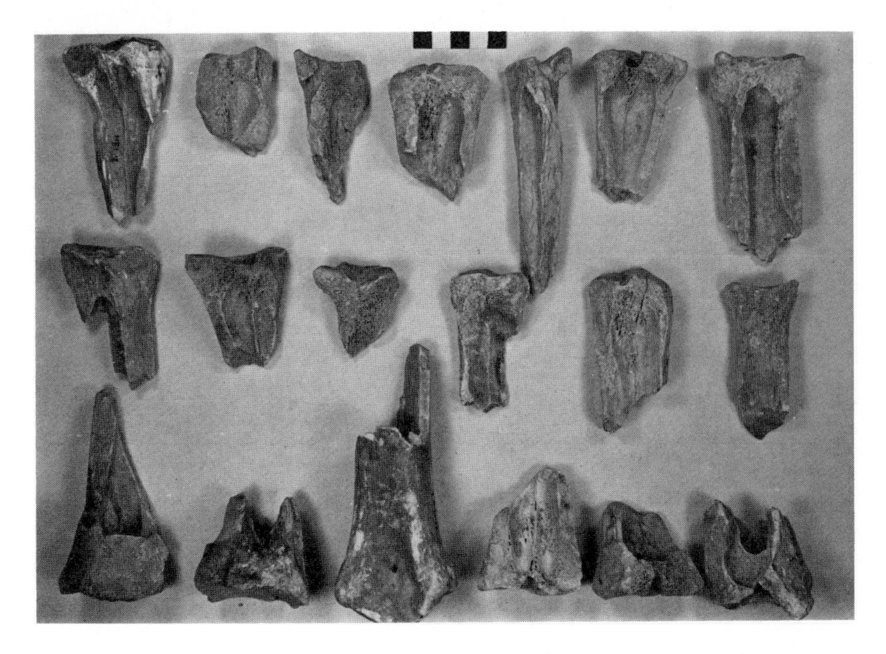

Figure 4.28 Split proximal metatarsals of eland (*Taurotragus*).

commonly unfractured, and if broken they are most commonly split into halves representing weathering fractures of young (largely unfused) animal bones. Put another way, the distal ends are not fragmented during processing, whereas the proximal end is extensively fragmented.

Examination of these breakage patterns strongly supports the reconstruction that the proximal ends were placed on an anvil and smashed with a fairly heavy hammer. I have observed this method of breakage among the Nunamiut Eskimo, particularly in hunting camps where the lower legs were only partially skinned prior to breaking for marrow. Men commonly place the unskinned lower leg next to the fire. There the hair is singed, the exposed skin is charred, and of course the marrow is warmed in the bone. The smoking lower leg is then taken from the fire, and the charred skin is cut and scraped back from the proximal end of the metapodial. Afterward, the cleaned proximal end is then placed on an anvil and hit with a heavy blow. The segments of the proximal end are then removed, and the split segments of shaft are peeled back, because they remain attached by periostium near their distal ends. The distal articular end remains encased in skin and unmodified by the process. Figure 4.29 shows the split and "peeled back" segments of two metatarsals with the splinters remaining attached to the distal ends, as they were abandoned on a Nunamiut hunting stand. The

Figure 4.29 Split metapodials remaining in a Nunamiut Eskimo hunting stand.

observations on burning (see Table 4.30) are completely consistent with this reconstruction of the context in which breakage of the metapodials is likely to have occurred.

METATARSALS AND PHALANGES

The breakage of the metatarsals is very similar to that of the metacarpals, differing only in the numbers of splinters normally generated when the proximal end is impacted. This is, of course a function of the difference in shape of the two metapodials. The split-shaft breakage of the metatarsals is well illustrated for the bushbuck bones shown in Figure 4.29. The splitting

TABLE 4.29

Breakage Patterning of First Phalanges[a]

	T. scriptus			Taurotragus			Pelorovis		
First phalanges	*N (1)*	*MNE (2)*	*% (3)*	*N (4)*	*MNE (5)*	*% (6)*	*N (7)*	*MNE (8)*	*% (9)*
Complete	7	7	.41	16	16	.14	4	4	.17
Proximal end	2	2	.12	32	32	.28	1	1	.04
Distal end	13	13	.76	63	63	.56	9	9	.38
½ proximal end	8	4	.24	68	34	.30	18	9	.38
½ distal end	7	3.5	.21	69	34.5	.31	21	10.5	.45
Total MNE		17.0			113.0			23.5	
Total MAU		2.13			14.13			2.94	

[a] MNE, minimal number of elements; MAU, minimal animal units (formerly called MNI).

Figure 4.30 Split phalanges of eland (*Taurotragus*).

of the bone by impacting on anvils, which appears to have been the processing strategies for the metapodials, is illustrated in a very clear fashion for phalanges. A cleaned and disarticulated phalange was seated on an anvil, ventral side down, and impacted on the ventral surface of the proximal end with a hammer. Small phalanges seem to have been supported by holding the distal end while impacting the proximal end. Larger bones from larger animals seem to have been less often held or supported, and impact was more commonly rendered in the center of the shaft (Table 4.29).

The fractured parts resulting from this inferred strategy are longitudinally split phalanges (Figure 4.30), longitudinally split articular ends, as well as complete proximal and distal articular ends with transverse or oblique breakage through the shaft.

Burning

A relatively small percentage of bones exhibited evidence of burning. In most cases the heat had not been sufficient to calcine the bone. Smudging or minor carbonizing of bone surfaces was the most common form noted (see Binford 1963).

With regard to the problem of whether animal gnawing had occurred prior to the introduction of parts to the site or after the site had been abandoned by hominids, there was one very interesting piece—a proximal femur of *Tragelaphus strepsiceros* (kudu) that had been gnawed by animals prior to having been burned. Of course it could be argued that the carbonizing was accidental and occurred by virtue of a fire kindled on top of a previously abandoned and scavenged bone. While not impossible, this seems unlikely in light of the frequency distribution of burning relative to anatomical-part frequencies. Table 4.30 summarizes the frequencies of burned fragments among those tabulated from the Klasies River Mouth site. Although it is clear that evidence of burning is relatively infrequent on the bones from the Klasies River Mouth site, it is equally clear that it is not random, nor does it covary with such properties as numbers of bone elements (see anatomical-part frequency data in Table 3.05). The two parts most consistently burned were fragments of occipital condyle and fragments of proximal femur.

I had previously observed a high incidence of burned occipital condyles among Mousterian faunal assemblages from France. In that case, parts of the skull (fragments not separated or counted regularly by Klein) such as the mastoid process, tended to covary with frequencies of burned occipital condyles and simple skull fragments. In the Mousterian materials, upper teeth were also frequently burned, tending to covary with occipital condyles and mastoid processes. This burning pattern results from the roasting of heads minus the mandible—the most common custom among the Nunamiut Eskimo, who roast heads in hunting camps and in and around kill-processing camps. There the mandible and attached tongue were commonly removed for transport to the village, whereas the cranium was considered of very marginal utility as regards transport inventments. For the latter reason this otherwise marginal part was frequently consumed in the field, not transported to residential sites. In cases where it was carried to living sites, the tongue and mandible would have been taken to the site early in the transport sequence, while the head would have been introduced later, perhaps as part of the last load or trip to the residence from the kill. This means that usually the mandible would have been romoved prior to the cranium's transport to the living site.

As mentioned earlier, a similar burning pattern was noted for the Mousterian from the site of Combe Grenal. This pattern was characteristic of red deer recovered from Würm I deposits and for reindeer recovered from Würm II deposits. In marked contrast, however, were the burned bone parts from red deer and horse (*Equus*), recovered from Würm II deposits. The latter two species exhibited no burning of occipital condyles, mastoid processes, and so forth, but instead had consistent burning of mandibles and mandibular teeth. This pattern is believed to result when the complete head is roasted with attached mandible. This is the form in which heads are most commonly roasted among the Kalahari Bushmen today and in the recent past (see Marshall 1976:89–91).

Figures 4.31 and 4.32 illustrate a roan antelope head being prepared for roasting among the Nyae Nyae Bushmen camped at Gautscha pan, Namibia. It is being roasted complete with mandible in a prepared roasting pit as shown in Figure 4.32. This procedure would most likely result in the burning of the margins of the mastoid process, and the occipital condyles would be less likely to rest directly on the hot coals. I have similarly observed the Navajo roasting heads of sheep. The heads are cooked complete—that is, with attached mandible. It was reported to me that bones from Navajo sites where heads were roasted commonly yielded mandibles with burned and charred undersurfaces.

It is particularly interesting that the pattern ethnographically reported from the South African area in recent times is not the pattern observed on the Klasies fauna. Instead, a pattern common to some species from Mousterian sites in Glacial Europe known from the contemporary Eskimo is represented. How is this to be understood? I must admit that I do not know. However, certain characteristics made explicit in these data are clearly worth investigating in the future.

The single relevant factor common to the Navajo and the contemporary Bushmen in their methods of preparing large game is that the complete animal is commonly used up in a very short time and generally at a single place. No long-term storage is involved, nor is there much specialized use of different anatomical parts, as was documented among the Nunamiut. For instance, it is commonly reported that contemporary Bushmen tend to reduce animals into four usable classes of materials—skin, head, meat, and bones. The skin and the head are prepared or cooked quite specifically. On the other hand, the meat is systematically cut from the bones into strips commonly referred to as *biltong,* and then the bones are processed for boiling. This means that the marrow is not eaten discretely but it is a contributor to the bone stew. In like fashion, the bone marrow is not eaten discretely among the Navajo. They tend to cook joints in a variety of stews, with the long bones butchered through the shaft.

Clearly, the pattern under discussion is not represented at Klasies River Mouth. I would not expect this, because there is no reason to anticipate the use of boiling strategies in the MSA of South Africa. Yet there is similarly no reason to anticipate such strategies in the Würm II Mousterian of south-central France, where I have observed the "modern" burning pattern! Obviously, then, conditioning factors were at work that we have not yet isolated. In the case of the pattern actually observed at Klasies River Mouth, it is most analogous to the pattern observed among the Nunamiut Eskimo, with the exception of the high frequencies of burned femur. The property

TABLE 4.30

Frequencies of Burned-Bone Fragments, Klasies River Mouth Fauna[a,b]

	Animal body-size classes										*Totals*	
	I		*II*		*III*		*IV*		*V*			
	No.	%	*No.*	%	*No.*	%	*No.*	%	*No.*	%	*No.*	%
Horn												
Occipital condyle	1	.25	2	.22	3	.17	4	.15	1	.50	11	.22
Mandible												
Atlas												
Axis												
Cervical Vertibrae												
Thoracic Vertebrae												
Lumbar Vertebrae												
Pelvis												
Scapula												
Proximal Humerus												
Distal Humerus												
Proximal radiocubitis					1	.08	3	.13	1	.09	5	.07

Distal radiocubitus												
Carpals												
Proximal metacarpal												
Distal Metacarpal												
Proximal femur	4	.22	2	.18	1	.07	4	.16	1	.20	12	.16
Distal femur			1	.07	1	.14	3	.21			5	.09
Proximal tibia												
Distal tibia	2	.09	1	.05								
Tarsals												
Astragalus												
Calcaneus							4	.10	1	.05	5	.04
Proximal metatarsal					1	.07	1	.03	1	.07	3	.03
Distal metatarsal					2	.14	1	.04	2	.13	5	.06
Phalanges												
First			2	.07			3	.02	1	.04	4	.01
Second												
Third									2	.07	2	.01

[a] Teeth were not examined for this property.

[b] Percentage is calculated as the proportion of the elements in the designated anatomical class which showed evidence of having been burned.

Figure 4.31 Roan antelope head prior to being prepared for roasting by Nyae Nyae Bushmen.

most characteristic of the Nunamiut system is that parts are differentially treated.

> In the Nunamiut case anatomical parts are differentially evaluated and this scale of evaluation is mapped onto different places and times. . . . On the other hand, the !Kung most certainly have some similar understanding of the differential utility of anatomical parts but this is mapped onto persons differentially evaluated in such terms as kinship association. (Binford 1978:133)

Almost certainly the burned lower-leg parts in the Klasies River Mouth sample represent bones warmed before being broken for bone marrow. This differential treatment is also noted among the Nunamiut. Similarly, the roasting of the head with the mandible removed also represents a differential treatment of the two head parts. The burning on the femur is most likely produced when a complete unskinned leg is roasted in an ash-roasting feature.

This type of differential treatment is foreshadowed at Klasies River Mouth in the differential treatment of animals of different body size, and in the clearly differential treatment of parts in butchering; for example, hacking on upper-limb bones and cutting on lower-limb bones. This Klasies pattern of burning is present in the Mousterian but only among some animals species, whereas others show different patterns of burning.

Figure 4.32 Roasting pit being prepared by Nyae Nyae Bushmen for cooking a roan antelope head.

This diversity, correlated either anatomically or with size-variable species, could inform us about different procurement strategies and, hence, different parts of varying utility available as a result of the execution of different procurement strategies. Scavenging would yield one set of anatomical parts from a range of animal species, whereas hunting or trapping would yield a different assortment of parts available. Such input diversity and variability may be manifest in processing differentiation that mimics some of the diversity noted in the Nunamiut case, deriving from differences in the time and space utility of parts obtained under essentially uniform strategies. This area for future research, if developed well, could be most informative regarding the organization of early man's subsistence strategies.

Interpretation of Patterning

Animal Gnawing

I have reported animal gnawing at Klasies River Mouth under the assumption that the gnawing had occurred prior to the recovery of the anatomical parts by the hominids, and prior to the introduction of bones into the site. I delayed the discussion of the possibility that the bones had been introduced into the site and subsequently scavenged after man had abandoned the Klasies rockshelter until the data from all the anatomical parts had been presented individually. This delay was simply to make possible the citation of the overall patterning in the gnawing as part of an argument that, in fact, the bones had been gnawed prior to their introduction to the Klasies site.

Several studies of prey carcasses (including natural deaths that had been scavenged) have been made (Binford 1981; Hill 1975). Richardson (1980b) provides a fair picture of the frequencies of parts as well as the condition of parts on kill–death sites after nonhuman predator–scavengers have fed on the carcasses. Table 4.31 combines a summary of the data reported by Richardson, from his study of scavengers and the ungulate carcasses remaining after they had completed their feeding with data from Klasies and an Anaktuvuk dog yard. Columns 1–3 present inventories of parts remaining at the kill site as well as the frequency of those remaining that were damaged by the scavengers. Column 3 shows the percentage of parts remaining that were damaged. Column 4 summarizes the percentage of bones exhibiting gnawing among the bones of animals in size class IV from Klasies River (eland-sized animals). Columns 5 and 6 display the percentage of surviving bones that were damaged during feeding by lions (Column 5) and hyaenas (Column 6), as observed by Richardson.

There are several facts of interest here. It should be pointed out that there is a considerable difference in the pattern of damaged bone remaining at a hyaena kill–feeding location (Column 6) as opposed to that seen on lion kills (Column 5). There is much greater damage on the long bones when hyaenas have been feeding. Similarly, there is greater damage on the skull and horns. (Richardson did not note the former, but I observed this in the Nossob River area). These observations lend some support to the views of several taphonomists (e.g., Haynes 1982) that we will eventually be able to recognize the species responsible for bone accumulations from the distinctive feeding patterns of each.

While these observations are interesting, other points of extreme interest are summarized in Figure 4.33, in which the percentages of bones in the various anatomical classes from Klasies that exhibited gnawing are plotted against the percentages of damaged bones in similar anatomical classes observed by Richardson at known sites of animal predation and scavenging. For the parts of the axial skeleton plotted on Figure 4.33A, it is obvious that there is a strong positive correlation between the two data sets. This means that the relative frequency of gnawing observed among different anatomical parts at the Klasies site is essentially the same as the relative frequency of animal-damaged parts remaining at carcasses after nonhuman predator–scavengers had finished feeding.

I consider this to be almost perfect proof that the parts at Klasies River Mouth were selected from a population of anatomical segments previously gnawed and damaged by predator–scavengers, such as lions and hyaenas. If a scavenger had gnawed on the bones after they had been introduced into the site at Klasies, I see no way that the patterns of gnawing would mimic patterns that are essentially conditioned by the sequence of dismemberment and feeding tactics characteristic of animals devouring a complete carcass. When scavenging an abandoned human site, the scavengers would have direct access to already dismembered body parts as well as parts that had been previously exploited; that is, the distribution of meat and muscle characteristic of a whole animal would have already been modified and/or removed before the scavenger arrived. In addition, the actual population of parts present would be biased in favor of some, whereas others would be underrepresented or absent. These facts should ensure that the pattern of gnawing would be different than when a complete carcass was being devoured or when a carcass's population of bones in different stages of dismemberment and prior use was being scavenged. The simple facts of access to certain parts would be vastly different for the scavenger arriving at an abandoned human site. Similarly, the parts that would be attractive in terms of food would be very different than when the animal faced a complete carcass.

As a partial control on the gnawing and damage to parts when a feeding animal has access to already dismembered parts, I studied a sample of bones collected from a dog-feeding area at Anaktuvuk Pass (see Binford and Bertram 1977:80, Sample I). The dogs tethered in this area had been fed anatomical segments that were secondarily butchered at the time of feeding into meal-size units for each dog. The only difference between the access to bone and what might characterize a scavenger opportunistically feeding across an abandoned human site is that there was something edible on all of the parts given to the dogs—which of course would not have been true for the parts remaining at a human site at the time of abandonment. This former fact is perhaps reflected in the very high frequencies of animal-damaged bones recovered from the dog areas (see Table 4.31, Column 7). The percentages of damaged bones even exceeded the damage done by feeding hyaena (see Table 4.31, Column 6) in nearly all instances.

The striking difference in the dog-yard data, however, is the equitable damage seen on all bones as compared with the damage patterns seen by Richardson at his scavenged carcasses. In the latter case, the bones of the appendicular skeleton were generally less damaged than were the bones of the axial skeleton. In a similar manner, the bones of the upper limbs were

TABLE 4.31

Richardson's Scavenging Data Compared with Klasies Large Bovids and Anaktuvuk Dog Yard

	Richardson's data[a]			Klasies bovids IV	Richardson's data[b] Percent gnawed		Anaktuvuk[c]
	Surviving MAU (1)	Damaged MAU (2)	Column 2 / Column 1 (3)	% of total MAU gnawed (4)	Lion (5)	Hyaena (6)	% Gnawed (7)
Horn	15.5	8.5	55.0	14.0	0	0	54.0
Skull	0	0	0	0	58.0	70.0	50.0
Mandible	15.5	8.5	55.0	9.0	44.0	25.0	66.0
Atlas–axis	14.0	4.5	32.0	0	20.0	25.0	75.0
Cervical vertebrae	14.5	7.5	52.0	8.0	56.0	47.0	82.0
Thoracic vertebrae	11.0	8.0	72.0	20.0	50.0	49.0	71.0
Lumbar vertebrae	11.5	7.5	65.0	29.0	92.0	56.0	0
Sacrum	15.0	10.0	67.0	25.0	62.0	17.0	?
Innominate	15.0	12.5	83.0	35.0	76.0	76.0	58.0
Scapula	12.5	10.0	80.0	26.0	40.0	28.0	73.0
Proximal humerus	11.0	5.5	50.0	29.0	8.0	45.0	0

Distal humerus	11.5	.5	4.0	0	0	23.0	70.0
Proximal radiocu-bitus	9.5	1.0	11.0	5.0	0	31.0	90.0
Distal radiocubitus	10.5	1.0	10.0	0	0	0	75.0
Carpals	0	0	0	0	0	0	26.0
Proximal metacarpal	8.0	0	0	8.0	0	0	61.0
Distal metacarpal	7.5	1.0	13.0	8.0	3.0	0	82.0
Proximal femur	9.0	3.3	37.0	28.0	8.0	14.0	75.0
Distal femur	9.5	2.0	21.0	15.0	0	64.0	75.0
Proximal tibia	11.0	2.0	18.0	29.0	0	32.0	84.0
Distal tibia	11.0	0	0	10.0	3.0	0	64.0
Tarsals	0	0	0	0	0	0	21.0
Astragalus	6.5	0	0	4.0	0	0	41.0
Calcaneus	10.5	0	0	10.0	0	0	53.0
Proximal metatarsal	10.5	0	0	3.0	0	0	65.0
Distal metatarsal	11.0	.5	5.0	4.0	3.0	2.0	82.0
Phalanges	2.5	0	0	3.0	0	0	13.0

[a] 1980b: Figure 39.
[b] 1980b: Figures 33 and 34.
[c] Binford and Bertram (1977:80, Table 3.1, Column IV).

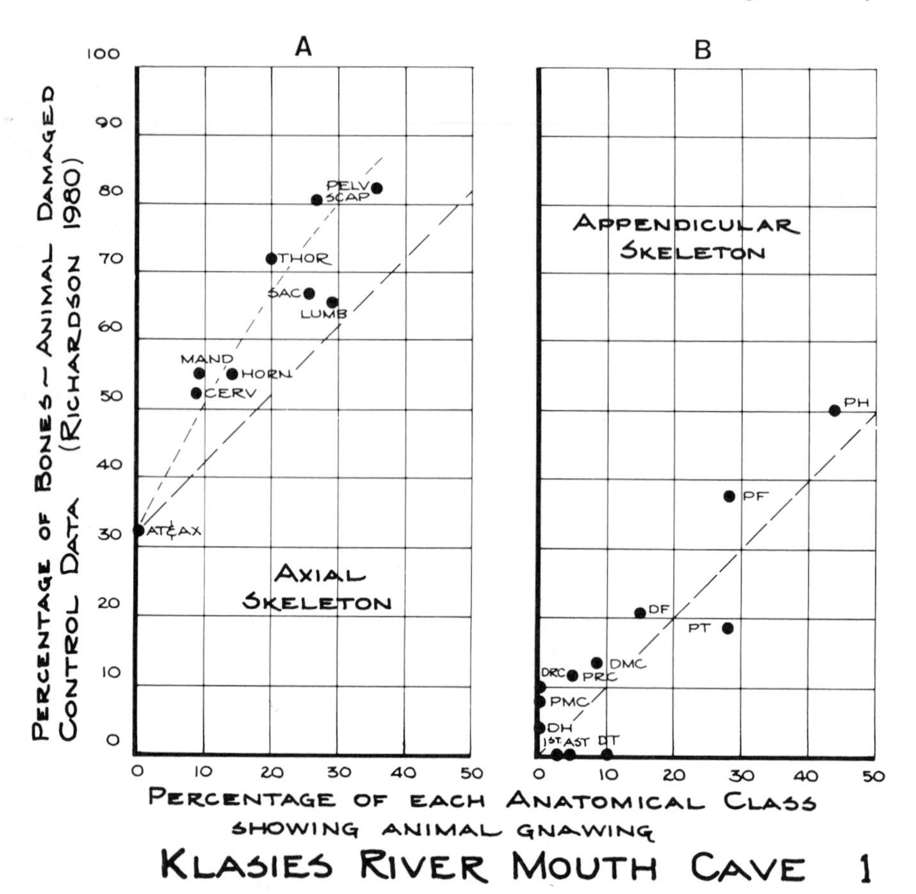

Figure 4.33 Comparison between control data (Richardson 1980) and the Klasies frequencies for animal gnawing on (A) axial and (B) appendicular skeletons. Long-dashed lines (in A and B) indicate a relation of identity between cases; short-dashed line (in A) shows actual relationship. AST, astragalus; AT, atlas; AX, axis; CERV, cervical vertebrae; DF, distal femur; DH, distal humens; DMC, distal metacarpal; DRC, distal radiocubitus; DT, distal tibia; HORN, horn; LUMB, lumbar vertebrae; MAND, mandible; PELV, pelvis; PF, proximal femur; PH, proximal humerus; PMC, proximal metacarpal; PRC, proximal radiocubitus; PT, proximal tibia; SAC, sacrum; SCAP, scapula; THOR, thoracic vertebrae; 1st, first phalange.

more damaged than were the bones of the lower limbs. Inspection of the data from the dog yards shows no such bias either with respect to the axial skeleton or the upper limbs. This is to be referred to the different patterns of access that the dogs experienced relative to the animals feeding at carcasses. The dogs can be assumed to have been hungry each time they were fed. The parts given to them were all edible, but in each case only the particular parts

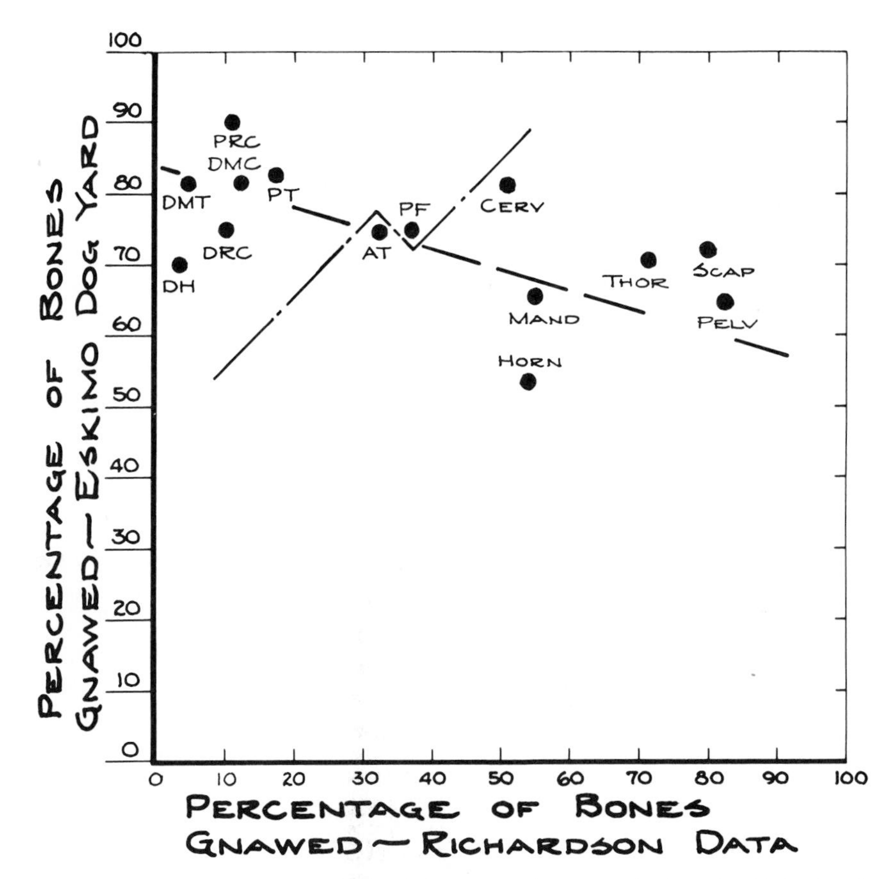

Figure 4.34 Percentages of animal destruction by anatomical part for Richardson's (1980) control data and a Nunamiut dog yard. AT, atlas; CERV, cervical vertebrae; DH, distal humerus; DMC, distal metacarpal; DMT, distal metatarsal; HORN, horn; MAND, mandible; PELV, pelvis; PF, proximal femur; PRC, proximal radiocubitus; PT, proximal tibia; SCAP, scapula; THOR, thoracic vertebrae.

given to the dog were accessible at each discrete feeding. This means that the dog could not pass up some parts in favor of others as would be the case of animals feeding at a carcass in which all parts were simultaneously available and edible (at least to some extent).

These differences between (1) the variability in the parts available and their food utility, and (2) the accessibility of parts by virtue of their separation as opposed to their incorporation into the functional anatomy of an animal, certainly condition the feeding patterns and intensities by the consuming animal. This is well illustrated in Figure 4.34, in which the percentages of bones present exhibiting gnawing or tooth scoring are plotted for

Richardson's data from known scavenged carcasses and for the bones recovered from the Eskimo dog yard.

There is a generalized negative or inverse relationship between the dog yard and the scavenged carcass data. I think this to be understood in terms of access by gnawing animals to bones. A dog fed a lower limb and several ribs devours what he has. The less actual food available, the more the dog devours the part to which he has access. This means that those dogs receiving a meal of marginal parts chew on them more intensely than does a better-fed dog receiving a greater bulk of food.

I do not want to place too much emphasis on these data. I introduce them solely to warrant the view that gnawing of bones is a function of the relationships between hunger and the accessibility of food. As hunger is abated, parts of marginal utility are ignored (the situation at carcasses). On the other hand, when hunger is present and the foods accessible are of minimal utility, the feeding animal will simply go after what is available (the yard case). Given such an understanding, it should be very clear that an animal scavenging an abandoned living site would address the parts that happened to be abandoned with some edible remnants. It is likely that these would be the parts not heavily processed or consistently used by the humans. I think it should also be clear that such parts are not likely to have been the choice meat-yielding parts in the anatomy, or the parts that represent the identical feeding biases of a predatory animal feeding on a complete carcass or a carcass sequentially scavenged. Referring again to Figure 4.33A, we see that the relative frequency of gnawed bones from the axial skeleton are the same as noted at carcasses known to have been killed and/or fed upon by nonhuman predator–scavengers. The major difference between the two sets of data rests in the fact that for every gnawed bone in the Klasies sample, there were three and a half to four such bones showing gnawing at the carcasses observed by Richardson. Given that the axial-skeleton parts essentially yield only meat as an edible product, it is not surprising that, if the hominids were recovering meat-yielding parts from previously killed and scavenged carcasses, they would select those that had not been heavily devoured already; that is, they would choose carcasses and and parts from carcasses that still had usable meat—hence carcasses below the mean as far as prior animal consumption was concerned.

I turn now to what I consider to be even stronger evidence than that from the axial skeleton—the data from the appendicular skeleton, as summarized in Figure 4.33B. Like the axial skeleton, there is a clear positive relationship between the relative frequencies of gnawing on different anatomical parts from Klasies River Mouth and such frequencies observed at carcasses by Richardson. However, although the relationship is as strong, it is not the same as was observed between the two data sets for the axial skeleton. For the legs, the relationship describes a line that is linear, with the intercept of the line at the origin of the graph. The slope of the line is essentially directly proportional, so that for each rise on the Y axis there is an identical increase on the X axis. This, then, means that the data from Klasies River is identical in all ways to the data from Richardson's observations on carcasses for parts of the appendicular skeleton. It will be recalled that this was not the case for parts of the axial skeleton. The line described by the correlation between the two data sets in the latter case intercepted the Y axis between a value of 30 and 40 and rose at an angle, so that for every unit increase in Y there was only about 75% of a unit increase in X. Stated simply, this means that the axial versus appendicular skeleton appears to have been an independently sampled segment from a single parent population. Both, however, were selected from an original population previously modified by animal gnawing in a manner essentially identical to the population summarized by Richardson. Parts of the appendicular skeleton selected were chosen for transport to Klasies River Mouth in terms of the degree that animals had not previously consumed the parts selected, a strategy completely consistent with a scavenger seeking meat.

On the other hand, the patterns of gnawing on the limb bones at Klasies River are nearly identical to the patterns of gnawing observed by Richardson at animal-scavenged carcasses. This implies that the prior consumption by predator–scavenging animals had not affected the utility for the hominids of the lower legs. It should be clear that the edible material being sought by the hominids was not meat but bone marrow. The latter is rarely consumed at kill–death sites by animals.

Figure 4.35 illustrates the leg of a springbok (*Antidorcas marsupialis*) that was observed being eaten by a leopard in the Nossob River Valley. This bone was all that remained hanging in the tree approximately 7 months after the observations on the leopard feeding. What is clear is that the leopard fractured through the humerus shaft (some tooth scoring is visible on the broken section of the humeral shaft), but the joints of the leg below the center of the humeral shaft were left untouched by the leopard. This pattern is repeated over and over again in which the lower limbs remaining at kill–death sites with their marrow yielding bones generally left untouched by predator–scavengers. (See Brain 1982:25, Figures 19 and 21, as well as p. 105, Figures 103 and 104.) This means that there are few limitations on the recovery of marrow-yielding lower limbs by hominids between death and when dangerous putrification or desiccation sets in. Even heavy scavenging by vultures need not render this resource unattractive to hominid scavengers. Because this resource could be used by hominids after other predator–scavengers had essentially finished exploiting a carcass, the pattern of large-carnivore–scavenger tooth-marks on limb bones in the Klasies

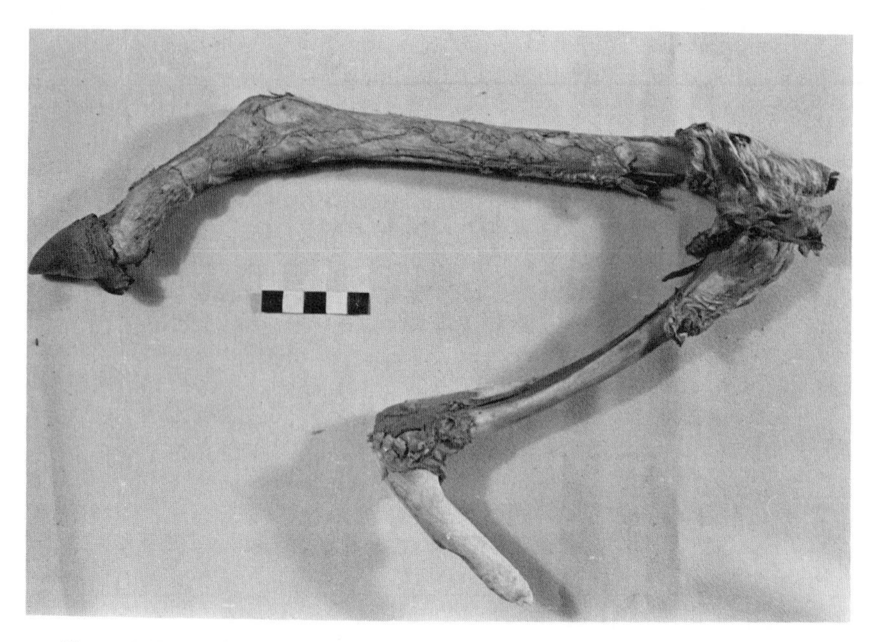

Figure 4.35 Modern springbok (*Antidorcas*) recovered from a leopard kill in the Nossob Valley, South Africa.

sample is identical to the pattern noted at kill–death sites after the carcasses had been exhausted by the predator–scavengers observed by Richardson.

The fact that tooth marking on the axial skeleton shows correlated but scalar differences relative to the control data of Richardson, whereas the bones of the limbs show a direct and near-identical pattern in both correlational form and scale, appears to me to be almost unassailable evidence that the tooth scoring occurred at kill–carcasses prior to the recovery of parts in a differential manner (one strategy for meat-yielding parts and another for marrow-yielding parts) by the hominids. No animal scavenging an abandoned hominid site would be confronted with such a complicated pattern of the parts that man just happened to leave lying around in identical ways each time the site was accessible to scavenging animals!

I should point out that to support the postabandonment argument, we would also have to model a situation in which each time man left his living site, the only bones lying around that were attractive to scavengers were from large animals (see the consistent pattern in essentially all the tables previously presented where gnawing is absent on the bones of smaller animals).

There is still an additional set of facts that I have not thus far discussed: the data from seal bones. I had the opportunity to study the bones of seals that were recovered from the Klasies River Mouth. Most of these bones were from relatively large and mature seals, and on them a consistent pattern of cut marks was observed that showed conclusively that these animals were butchered by man. On the other hand, of all the bones, only seven showed any animal gnawing. In four cases there were large tooth punctures through the phalanges, or minor gnawing on the distal ends of the humerus or on the radius; but no other bones were gnawed. All gnawing was concentrated on the bones of the front flipper. I see no reason why a scavenger eating opportunistically across an abandoned surface would never have encountered edible seal bones from the meaty areas of mature seals previously butchered by hominids. Once again, we would have to imagine the hominids eating smaller ungulates and seals to the point of their being unattractive to scavengers, while regularly leaving choice parts of the larger ungulates with edible remnants! The most economical as well as the best-informed interpretation is simply that the hominids were systematically scavenging the carcasses of the larger ungulates. This scavenging generally took place after the other predator–scavengers in the area had completed their feeding at the carcass. Scavenging consisted of recovering parts with usable meat when possible, but the most regular practice was the recovery of marrow-yielding lower legs from such carcasses. These large animals were not hunted. Hunting or trapping—that is, a direct procurement strategy—seems restricted to relatively small-body-size ungulates, and, as argued earlier, primary consumption of these small animals was not generally conducted at the Klasies site.

Breakage

Bone breakage seems most clearly indicative of (1) extensive hominid processing of metapodials and phalanges from larger animals for bone marrow; (2) the regular processing of horn cores from the tragalaphines (*Taurotragus oryx* [eland]), *Tragalaphus strepsiceros* [greater kudu]), and *Tragalaphus scriptus* [bushbuck]) for the small morsels of edible matter contained in the horn-core sinuses; and (3) regular processing of mandibular margins of the larger animals for the small morsels of edible pulp recoverable from along the bases of the mandibular teeth. In addition to the breakage referable to hominid bone-processing, there was distinctive breakage of the vertebrae and ribs in particular, which is referable to ravaging by feeding animals.

Several points are very important regarding breakage. Perhaps the first and clearly the most provocative is that there was intense processing for what can only be considered the most marginal food—pulp from horn-core sinuses, and mandibles, as well as marrow from phalanges and metapodials. This processing was regularly carried out on parts recovered from the larger animals. Great labor investments were made by the hominids of Klasies River Mouth in recovering small amounts of marginal foods from parts originating in animals that, if exploited fresh or killed by the hominids, would have provided enormous quantities of high-quality foods accessible without major processing.

A conclusion seems inescapable to me: large quantities of high-quality foods were simply not available at the carcasses of these large animals. This conclusion is supported by observations on the utilization of the small bovids. It must be recalled that in terms of body-part frequencies, cut marks, and the absence of animal gnawing, it seemed clear that these small creatures were being exploited for meat. In addition to this inference, the pattern of meat utilization seemed to have been of a "gourmet" form, where only small and choice parts were regularly introduced. In like fashion, parts of marginal utility on the small animals were not generally introduced or processed. This clearly indicates that, at least when high-quality meat was available, the hominids consumed it and did not simultaneously process and consume parts of low or marginal value. Given that they regularly introduced parts of very low potential food value from the large animals, the choice meat was most likely not available. This view is supported by the breakage patterns on the ribs and vertebrae introduced from the large animals. It will be recalled that the regular breakage of the dorsal crests of the dorsal vertebral spines, coupled with the regular breakage of the ribs out some 5 cm (2 inches) from the proximal articular surfaces of the rib (see Figure 4.36), are both consistent with animal feeding strategies (evidenced by gnawing), and the feeding on at least partially stiff carcasses. It would appear that the introduction to the site of such parts, previously broken by animal feeding, clearly suggests that small, remnant, and at least partially dried strings and sections of tenderloin and adhering meat were the food targets remaining on these anatomical parts.

The breakage pattern referable to processing of parts is largely a feature of bones from the larger animals. Processing investments were in anatomical segments that could yield at best only small quantities of rather marginal foods. This pattern is consistent with the exploitation of large animal carcasses where prime foods were not present or had already been consumed. The latter condition is strongly indicated by the distinctive breakage patterns of the ribs and dorsal spines of the vertebrae (see Figures 4.9, 4.10, 4.11 4.36).

All in all, the pattern is clear: the hominids were introducing parts of small size from the already relatively dry and heavily exploited carcasses of large animals, and the small parts yield very small amounts of food of largely marginal value. Hunting of the large forms seems not to have been part of the hominids' food-getting strategy.

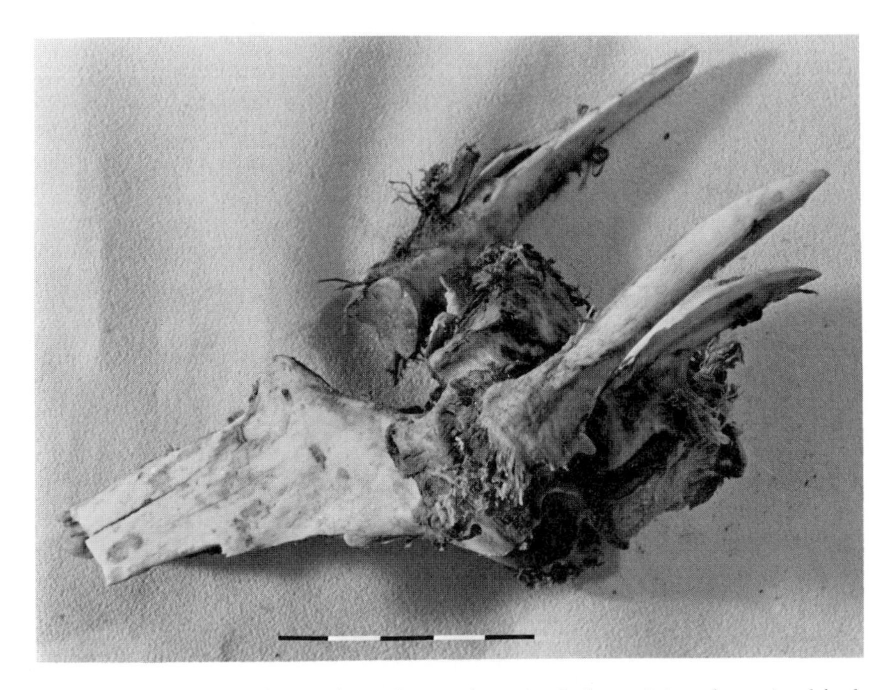

Figure 4.36 Typical unit of vertebrae and proximal rib surviving after animal feeding: modern wildebeest (*Connochaetes*) fed upon by hyaena in the Nossob Valley, South Africa.

Cut Marks and Evidence of Dismemberment

As has been pointed out throughout the descriptions of the body parts, at least three different formation conditions may be signaled by the position and character of cut marks: (1) we may recognize dismemberment as opposed to skinning or filleting of either muscle or sheathing for bony parts; (2) we may also recognize something of the state of the carcasses at time of dismemberment with regard to whether it is supple and fresh or stiff, and in an African context, desiccated and old; and (3) we may recognize something of the dismemberment and processing strategy employed.

AXIAL SKELETON

Data generated from observations on the axial skeleton signal unequivocally the fact that the small animals (small and small–medium bovids) were selected and processed in terms quite differently from the medium- through large-body-size animals. I have prepared Table 4.32 using the crite-

TABLE 4.32

Summary of Inflicted Marks by Size Category

	Bovid class																	
	I								IV								Combined dismem-berment cut and hack marks	
	Cut marks								Cut marks									
	Dismem-berment		Filleting		Hack marks		Gnaw marks		Dismem-berment		Filleting		Hack marks		Gnaw marks			
	No. (1)	% (2)	No. (3)	% (4)	No. (5)	% (6)	No. (7)	% (8)	No. (9)	% (10)	No. (11)	% (12)	No. (13)	% (14)	No. (15)	% (16)	No. (17)	% (18)
Horns	1	.03	0	0	0	0	1	.03	0	0	0	0	6	.15	3	.07	6	.15
Occipital condyle	4	.36	0	0	0	0	0	0	0	0	0	0	0	0	3	.08	0	0
Maxilla	0	0	0	0	0	0	0	0	4	.18	0	0	1	.05	0	0	5	.23
Mandible	3	.05	0	0	0	0	0	0	7	.09	0	0	8	.10	5	.06	15	.19
Atlas	4	.33	0	0	0	0	0	0	0	0	0	0	0	0	0	0	0	0
Axis	1	.05	0	0	0	0	0	0	0	0	0	0	0	0	0	0	0	0
Thoracic spines	0	0	28	.32			0	0	0	0	8	.12	3	.04	3	.04	3	.04

Thoracic vertebrae (centrum)	1	.01	0	0	0	0	3	.02	0	0	0	0	2	.01	21	.14	2	.01
Ribs	0	0	0	0	0	0	0	0	1	.04	8	.30	13	.48	6	.22	14	.52
Lumbar vertebrae	3	.04	12	.14	0	0	3	.04	1	.01	0	0	0	0	21	.18	1	.01
Sacrum	1	.06	4	.24	0	0	0	0	0	0	2	.20			5	.50	0	0
Pelvis	(	8	.18	)[a]					(	7	.17	)[a]	1	.02	11	.26	?	—
Proximal femur	5	.17	0	0	0	0	1	.03	4	.09	0	0	0	0	10	.22	4	.09
Distal femur	8	.10	0	0	0	0	0	0	11	.18	0	0	0	0	6	.10	11	.18
Proximal tibia	3	.13	0	0	0	0	0	0	7	.25	0	0	0	0	4	.14	7	.25
Distal tibia	9	.21	0	0	0	0	0	0	10	.11	0	0	0	0	4	.04	10	.11
Proximal metatarsal	0	0	0	0	0	0	0	0	13	.19	7	.10	9	.13	5	.07	22	.32
Distal metatarsal	0	0	0	0	0	0	0	0	5	.09	0	0	3	.05	1	.02	8	.14
Scapula	27	.18	3	.02	0	0	0	0	1	.01	6	.06	7	.07	19	.18	8	.08
Proximal humerus	0	0	0	0	0	0	1	.10	2	.15	0	0	0	0	4	.31	2	.15
Distal humerus	8	.10	0	0	0	0	0	0	11	.18	0	0	0	0	0	0	11	.18
Proximal radiocubitus	6	.26	0	0	0	0	3	.12	0	0	4	.09	1	.02	1	.02	1	.02
Distal radiocubitus	0	0	0	0	0	0	0	0	0	0	3	.08	0	0	0	0	0	
Proximal metacarpal	3	.19	0	0	0	0	0	0	3	.03	17	.21	4	.05	6	.08	7	.08
Distal metacarpal	0	0	0	0	0	0	1	.07	7	.09	3	.03	1	.01	4	.05	8	.10

[a] Grouped because of ambiguity.

ria advanced by some of the early students of cut marks (Guilday *et al.* 1962:63)—namely, that dismemberment marks are recognizable by repetitive placement and "that there was some anatomically dictated reason why a particular mark should occur at any given spot." This Table summarizes by anatomical part the frequencies of the cut, gnaw, and hack marks on the bones of the large animals (size classes III to V) and the small animals (size classes I and II) as described in the previous descriptive sections. These tabulations are broken down into columns summarizing marks judged to have been inflicted while dismembering animals, and marks judged to have been inflicted while filleting or skinning anatomical segments.

Figure 4.37 displays the percentages of bones marked during dismemberment from the Klasies River small animals, plotted against the percentages of marked bones from the control population (Table 3.3, Column 4) observed among the Nunamiut Eskimo during their spring hunting and meat-drying operations. It is clear that two parallel, positively correlated linear relationships are indicated. The upper line is described by horns; occipital condyles; atlas and axis vertebrae; the lumbar and thoracic vertebrae; and, from the front leg, the scapula, proximal radiocubitus, and proximal metacarpals; and from rear leg the proximal femur and distal tibia. These are all the parts that exhibit a strong positive correlation between the two cases. This means that, at least as far as these parts are concerned, the dismemberment strategies of the Nunamiut Eskimo and the hominids of Klasies River were essentially the same. The only significant difference is that slightly more marks were inflicted on most parts by the MSA hominids than by the modern Eskimo using metal knives.

The lower line in Figure 4.37 also describes a positive correlation between the Nunamiut data and the Klasies small bovids, but it differs in the fact that many more marks appeared on the Eskimo bones relative to marks appearing on the Klasies fauna. The suite of bones marked in a similar pattern, but generally less commonly than the Eskimo bones, were the pelvis, distal femur, proximal tibia, and the proximal and distal metatarsals. Ribs were less commonly marked at Klasies River, as were the proximal humerus, distal radiocubitus, and distal metacarpal.

I think it should be clear that butchering and dismemberment strategies are at least a partial function of the size of the animal being addressed. Large animals are generally segmented into more units than are very small animals. Put another way, there is a package size that is a generalized target of dismemberment. With very small animals this basic package size is achieved with less dismemberment than would be the case when butchering a large animal. This difference accounts for the independently distributed but similarly correlated patterns seen in Figure 4.37.

This split pattern has a number of interesting implications. The basic

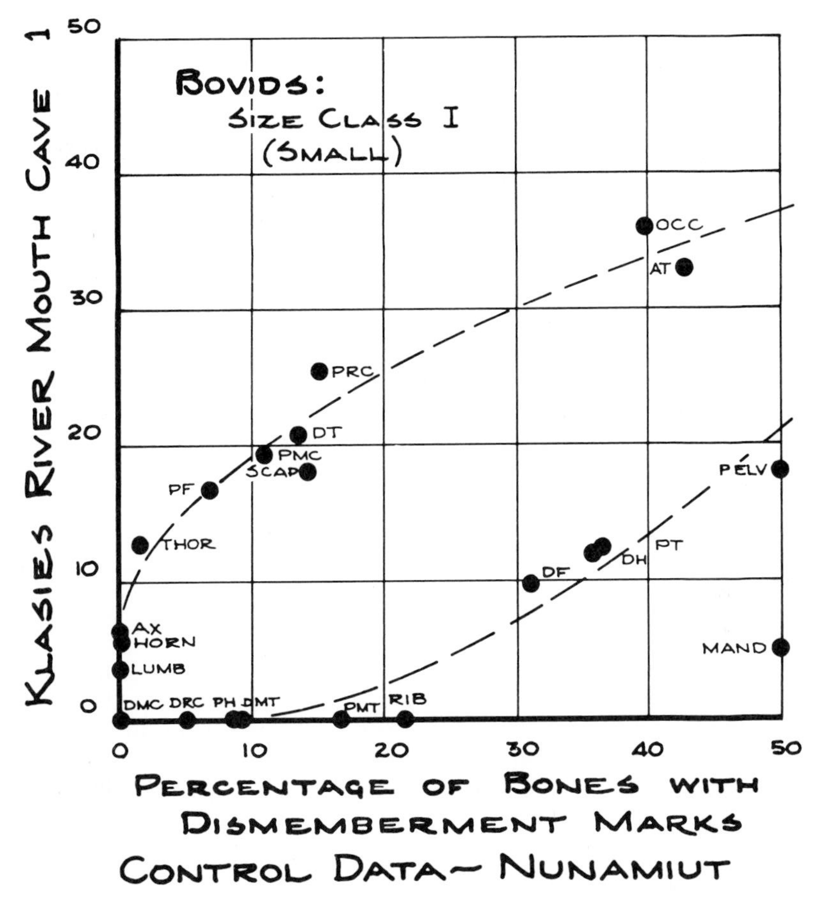

Figure 4.37 Comparison between dismemberment marks in a Nunamiut control population and from the Klasies River Mouth small-animal population. AT, atlas; AX, axis; DF, distal femur; DH, distal humerus; DMC, distal metacarpal; DMT, distal metatarsal; DRC, distal radiocubitus; DT, distal tibia; HORN, horn; LUMB, lumbar vertebrae; MAND, mandible; OCC, occipital condyle; PELV, pelvis; PF, proximal femur; PH, proximal humerus; PMC, proximal metacarpal; PMT, proximal metatarsal; PRC, proximal radiocubitus; PT, proximal tibia; RIB, rib; SCAP, scapula; THOR, thoracic vertebrae.

investment in dismemberment was similar between the Nunamiut Eskimo and the hominids butchering small antelope. Stated another way, the dismemberment tactics were similar relative to the basic features of the anatomy. The only real difference seems to be referable to the differences in body size of the animals being butchered. This is strong confirmation of the assumptions upon which inferences from dismemberment-mark frequency patterning are commonly based—namely, that these marks are anatomically clustered in terms of frequency, as a function of the relative investments (work) made by ancient butchers in dismembering carcasses in ways related to gaining usable access to foods that were differentially distributed on the skeletal framework. High investments would be correlated with acts of removing anatomical parts of high food-yield; hence such parts should be targets of removal more commonly than parts of low yield.

This overall pattern is realized in the dismemberment mark frequencies seen among the small mammals at Klasies River. There is little doubt that the dismemberment was performed in the context of strategies aimed at recovering usable meat from the carcasses of these small animals. The data from dismemberment marks are thus internally consistent with the inferences drawn from anatomical-part frequencies of the small-bovid classes: the hominids of Klasies River Mouth were exploiting the small bovids for meat, and they were generally obtaining the carcasses fresh—that is, prior to their becoming stiff. This condition is most consistent with their having been either hunted or trapped, or at least killed by the hominids. This view is supported by the near total lack of carnivore gnawing on the bones of the small bovids.

In terms of filleting marks (see Table 4.31)—marks inflicted during the removal of meat and/or skin from the bones—the pattern among the small bovid bones is very clear. These marks are exclusively present on the bones that yield most meat, and/or which are mechanically difficult to fillet: the thoracic spines (removal of the tenderloins), dorsal spines of the lumbar vertebrae (removal of the tenderloins), sacrum (removal of the tenderloins), and the scapula (removal of meat strips from the upper front quarters). Absence of such marks from parts of the rear legs is consistent with the patterns of burning in which rear legs appear to have been frequently cooked as complete units in their skins.

Filleting of both the tenderloin and the upper-front quarter may well betray the preparation of *biltong*, or air-dried stripped meat, for future consumption. If so, this suggests very short-term planning and a lack of sharing, because these were very small animals indeed. If sharing had been widespread, such small animals would likely have been shared out for consumption, with no surplus available for drying.

However, filleting out of the tenderloin and the scapula may not be related exclusively to preparation of biltong. These are both areas that cannot be butchered out because there is a protective layer of skin around the meat. Skin is only on one side of the scapula, and would be impossible to butcher out of the tenderloin area so as to encase it in skin. Filleting may be an alternative procedure in preparation for cooking when roasting in the skin is not feasible. In any event, all the data from anatomical parts, dis-

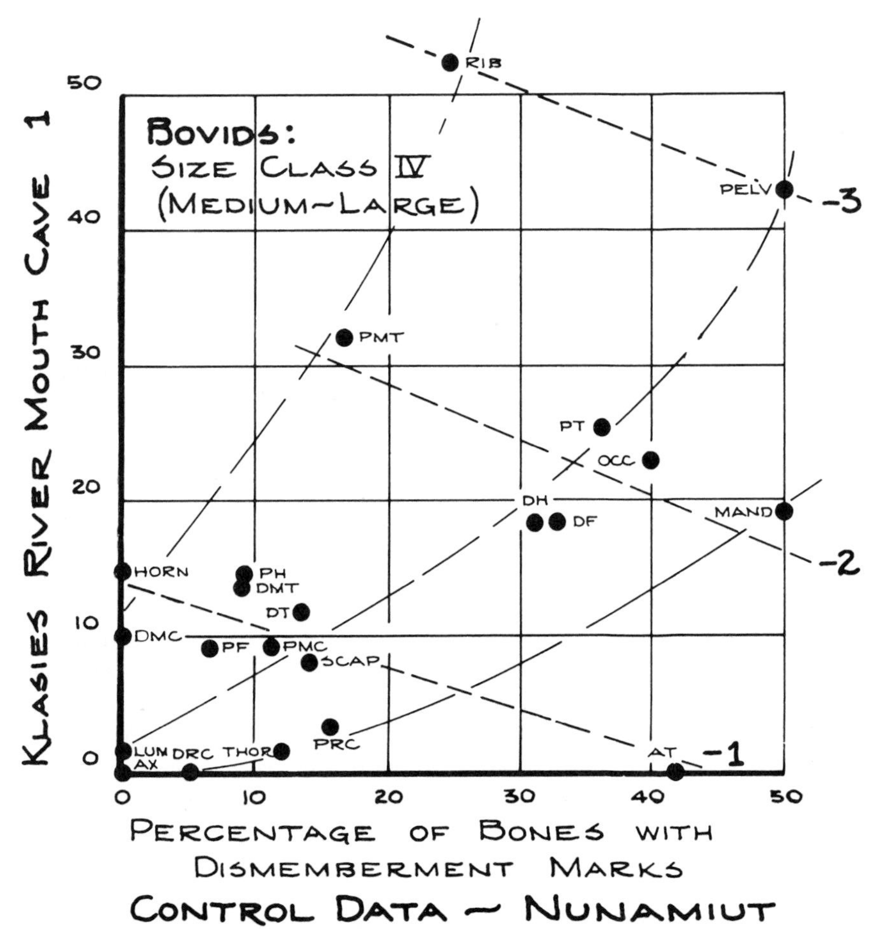

Figure 4.38 Comparison between dismemberment marks in a Nunamiut control population and from the Klasies River Mouth large-animal population. AT, atlas; AX, axis; DF, distal femur; DH, distal humerus; DMC, distal metacarpal; DMT, distal metatarsal; DRC, distal radiocubitus; DT, distal tibia; HORN, horn; LUMB, lumbar vertebrae; MAND, mandible; OCC, occipital condyle; PELV, pelvis; PF, proximal femur; PH, proximal humerus; PMC, proximal metacarpal; PMT, proximal metatarsal; PRC, proximal radiocubitus; PT, proximal tibia; RIB, rib; SCAP, scapula; THOR, thoracic vertebrae.

memberment marks, filleting marks, and animal gnawing point to the same conclusion: small animals were obtained fresh, butchered fresh, and primarily exploited for meat yields.

Turning to the data from the large animals, we obtain a very different picture. Figure 4.38 displays the relationships between the dismemberment marks (both cut and hack marks, Table 4.32, Column 17) and the control data on dismemberment from the Nunamiut Eskimo (Table 3.3, Column 3). Unlike the situation with the small animals, where there was a clear, correlated relationship between the Klasies materials and the control data, the pattern for the large mammals is spread all over the graph relative to the control data. Clearly, any attempt to fit these data would yield a strong indication of no relationship. However, there do appear to be some complicated groupings within the distribution, so that some sets of parts appear positively correlated (horn, proximal metatarsal, and ribs), although each is arrayed in a further grouping that appears to be negatively correlated (proximal metatarsal, proximal tibia, distal humerus, distal femur, occipital condyle, and mandible). This type of partitioned distribution is common when one or more additional factors are contributing to the patterning and these are not monitored. In short, there is a strong multidimensional set of determinants at work and the control data only account for a small proportion of the total variance. We may suspect that the frequencies of dismemberment marks are conditioned in this case by considerations other than simple variations in (1) body size and (2) differental proportions of meat on the skeletal framework of the large animals.

Given the assumption that the amount of inflicted marks should correspond to the amount of labor invested in dismembering, I would have to conclude that dismemberment of the large bovids and the dismemberment of the caribou used as a control set of facts reflecting dismemberment for meat were in terms of different goals. This "coping with other conditions" in the large-animal case is certainly implied in the earlier analysis of anatomical parts, and in the discussion of gnawing marks. Both of these characteristics were patterned so as to strongly suggest (1) biased selection of parts for marrow recovery, and (2) common recovery of large-bovid parts from carcasses previously ravaged by carnivores. Both of these suggestions are consistent with the lack of relationship between the large-animal dismemberment-mark frequencies and the control Eskimo data, in which the carcasses had been butchered fresh to recover maximum amounts of meat.

Still further differences between small and large animals were indicated by the cut marks themselves. It will be recalled that when there was evidence of butchering relatively stiff joints, this was exclusively a property of the large-bovid material. This alone suggests that the amount of labor, and hence numbers of inflicted marks, might be expected to vary with the degree of dismemberment difficulty, and not only with the numbers of dismembering acts as when fresh, supple carcasses are being addressed. Consistent with the degree of difficulty as a function of different carcass states as a conditioner was the fact that inflicted marks from heavy-handed chopping actions are common on the bones of the large animals. When appearing on limb bones, such marks almost certainly betray a dry and desiccated state.

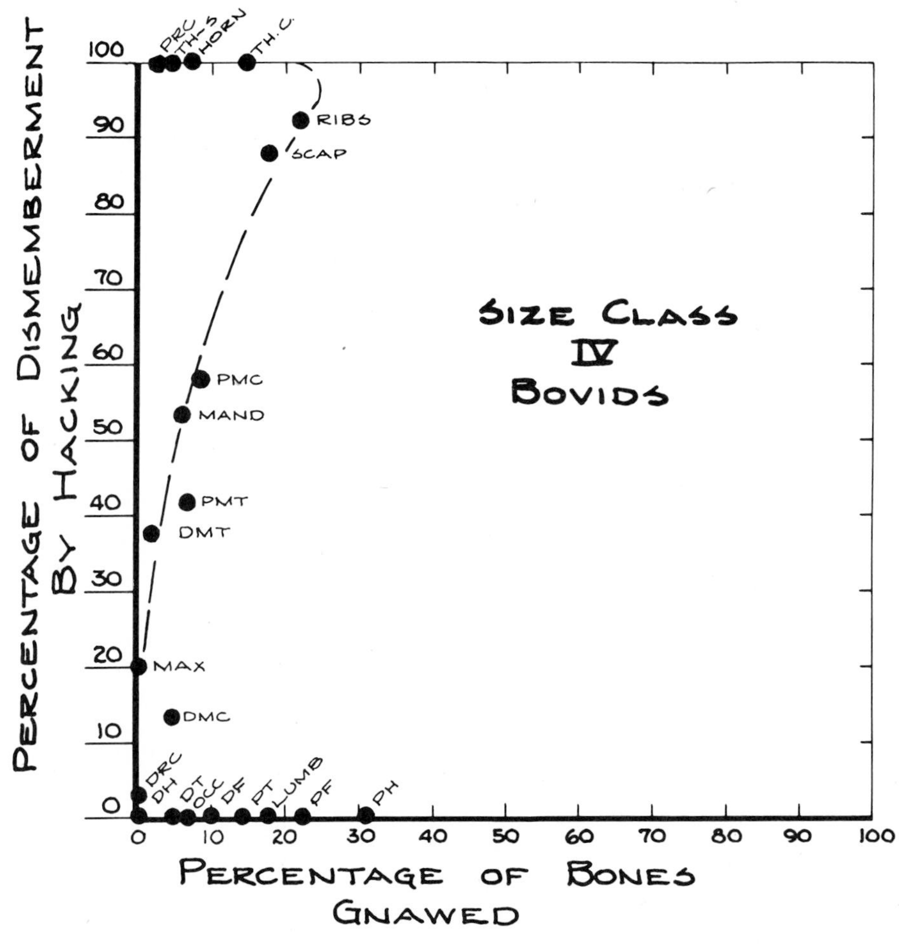

Figure 4.39 Comparison between frequencies of bones bearing chopping dismemberment marks and bones exhibiting gnawing marks. DF, distal femur; DH, distal humerus; DMC, distal metacarpal; DMT, distal metatarsal; DRC, distal radiocubitus; DT, distal tibia; HORN, horn; LUMB, lumbar vertebrae; MAND, mandible; MAX, maxilla; OCC, occipital condyle; PF, proximal femur; PH, proximal humerus; PMC, proximal metacarpal; PMT, proximal metatarsal; PRC, proximal radiocubitus; PT, proximal tibia; RIBS, ribs; SCAP, scapula; TH.C., thoracic centrum; TH-S, thoracic spine.

Figure 4.39 displays the relative percentage of dismemberment marks that were chopping or hack type marks, displayed against the percentage of the bones in each anatomical category exhibiting animal gnawing. A very interesting segregation results from plotting these properties. First, at the bottom of the graph are the parts on which slicing marks are rare or generally absent, and hacking–chopping marks are common. It should be noted that these are the parts that, with the exception of the mandible, were all grouped together when animal gnaw marks were plotted against the control data from Richardson's observations on known scavenged carcasses (see Figure 4.33). These parts were underrepresented in animal gnawing relative to the frequency with which it occurred on the controlled carcasses. It was suggested that this was to be understood in terms of biased selection by the hominids of those parts from carcasses that had not been heavily ravaged by other scavengers, because the only really usable material on these parts was meat. If other scavengers had already been at a carcass sufficiently prior to the arrival of hominids, there is likely to be little meat left on these parts. Therefore, when hominids did remove these parts, it is likely to have been at natural death sites where other predators had not yet had first shot at the carcass.

The data on dismemberment marks add another provocative piece of information. When these axial parts were returned, they had almost without exception been butchered by heavy-handed chopping tactics. This is likely to have been the case only if the carcass had been partially dried and stiff prior to dismemberment by the hominids. This is very strong confirmation of the view that these meat-yielding parts were recovered from death sites that hominids encountered while searching for food and not from kills that they themselves had made.

It should be noted that there is, in general, a very clean inverse curvilinear relationship between the frequency of gnawing and dismemberment marks of the cutting variety for the sacrum, ribs, scapula, proximal metacarpal, mandible, proximal metatarsal, distal metatarsal, and maxilla. Put another way, sacra, for instance, are apt to have been dismembered by hacking and also exhibit substantial gnawing marks, whereas maxillae are apt to be dismembered by cutting and to exhibit no gnawing marks. Mandibles and parts of the metapodials are all apt only infrequently to be moderately gnawed, yet to have a roughly equal chance of having been dismembered by either cutting or chopping. I return to this characteristic when discussing filleting.

Grouped at the bottom of the graph in Figure 4.39 are the proximal and distal humerus, proximal and distal tibia, and proximal and distal femur. These are the upper-limb bones that would normally yield the greatest amount of meat. The only heavy meat-yielding part not represented is the scapula, which, as we have seen, behaves with the axial skeleton both in terms of gnawing frequency (Figure 4.33), and incidence of chopping (Figure 4.39). This is consistent with the repeated observation that the scapula is one of the first bones to be dismembered by feeding animals. (See Binford 1981; Haynes 1982:271; Hill 1979:742; Richardson 1980a; and Shipman and Phillips-Conroy 1977.)

The meat-yielding bones clustered at the bottom of Figure 4.39 have been shown to exhibit animal gnawing in direct proportion to that observed by Richardson at control carcasses observed while being ravaged. Nevertheless, these parts are exclusively marked by slicing cuts, showing that dismemberment was only done with knifelike tools that are, as pointed out, only appropriate for cutting still-supple meat. That this pattern of dismemberment marks on upper-limb bones is relatively rare (see Table 4.32, Column 2), but when present the marks are "knife"-inflicted, suggests that recovery of meat from fresh carcasses was the goal. This pattern of meat recovery from high-yield parts of the upper legs appears as a gourmet strategy (see Binford 1978), which, as has been pointed out when discussing body parts from small animals, is inconsistent with the return of large quantities of available meat to a home base.

When faced with relatively unravaged carcasses of large animals (representing sometimes really vast quantities of meat), the hominids' response seems to have been to eat one's fill at the carcass and to return occasionally to the home site with a highly selected choice part. This pattern is consistent with the situation seen for the small bovids, for which transported meat-yielding parts were second-order parts. The implication is that first-order parts were consumed at the carcass, since it is likely that if unravaged, meat-yielding parts were available from the upper-front leg, rear-leg meat was also present at the carcass. On the other hand, among the large animals, consumption at the carcass would not necessarily result in first-choice parts being unavailable for transport. On the large-bovid bones, chopping and hacking was most common on parts of the axial skeleton and the scapula. These are parts that would be clearly second-order parts in a standard choice sequence. The chopping and hacking strongly indicate that these parts were stiff at the time of dismemberment. Interestingly, such parts were generally ignored when fresh meat was available, as is indicated by the slicing marks on the upper-limb bones, but were exploited when fresh meat was unavailable. On these parts carnivores have difficulty in stripping off all the adhering meat because of the irregular shapes of the bones. It is obvious, I think, that these parts were introduced from carcasses that had no substantial quantities of meat left to offer. The pattern of exploitation seems clear. When large quantities of meat were available, the hominids presumably fed at the carcass until full, and occasionally carried back to the Klasies site a few selected parts (a gourmet strategy), presumably abandoning the remaining meat at the carcass. More commonly, however, they encountered heavily ravaged carcass parts and, with some processing (soaking and/or breaking open the bones, mandibles, and some horn cores), these marginal parts could be made to yield some tidbits of food.

These tactics betray a very short-term planning and an almost "stimulus–response" structure of behavior. Such strategies are inconsistent with a model of provisioning the occupants of the site, unless the occupants were very small groups indeed, and did not intend to stay at the site much beyond one feeding interval. In either case, the implication is one of a pattern of feeding behavior very different from what would be expected of modern men behaving as hunters and gatherers.

An interesting additional fact is indicated by the distribution displayed in Figure 4.39. Distal ends of metapodials tend to have consistently more cut than hack marks, and the discrepancy is much greater for metacarpals than for metatarsals. Proximal metatarsals have over 60% of inflicted dismemberment marks made by cutting tools as opposed to 45% for metacarpals. Looking back on the data on anatomical-part frequencies (Tables 3.4 and 3.5), it will be recalled that, among the moderate-size animals, metacarpals were introduced to the site far more frequently than metatarsals, which is consistent with the model of hominids most often scavenging usable parts that were already dismembered from a carcass—that is, picking up already loose parts rather than actively dismembering carcasses.

The picture I get from reviewing the cut marks against a picture of the anatomical-part frequencies is that there is a bias in favor of (1) parts that require processing—that is, relatively large, fresh food packages in their own skin containers, which can be cooked, or parts that require filleting off the bones if they are to be cooked in a skin packet or hung over the flames—and (2) parts that require extra processing to recover the marrow (such as dry lower-limbs). This latter point is further amplified by the details from filleting marks.

Marks inflicted as part of filleting operations are perhaps the most revealing of any characteristics thus far discussed. Figure 4.40 displays the frequencies of filleting marks from the large bovids at Klasies River Mouth against the control data among the Nunamiut Eskimo. For the set of bones made up of the femur (filleting upper-rear leg), scapula (filleting upper-front leg), and thoracic spines (filleting the tenderloin), there is a positive correlation between the Klasies data and the control material, differing only in the gross frequency of inflicted marks (40–70% of the bones showing marks among the Nunamiut fauna known to have been filleted as opposed to only 0–10% of the same bones showing filleting marks at Klasies River Mouth). On the other hand, the bones most commonly showing filleting marks at Klasies River are bones that rarely, if ever, exhibit such marks in the control data. Particularly interesting in this regard are the ribs and the proximal metatarsals and metacarpals, as well as proximal and distal radiocubiti. These are all bones that have little adhering meat, and, in the case of the metapodials, no meat at all. It is a near certainty that the marks on the lower-limb bones were inflicted while skinning bones in preparation for marrow cracking.

In addition, there is the implication that when the bones were skinned in preparation to having their proximal articular ends smashed on an anvil,

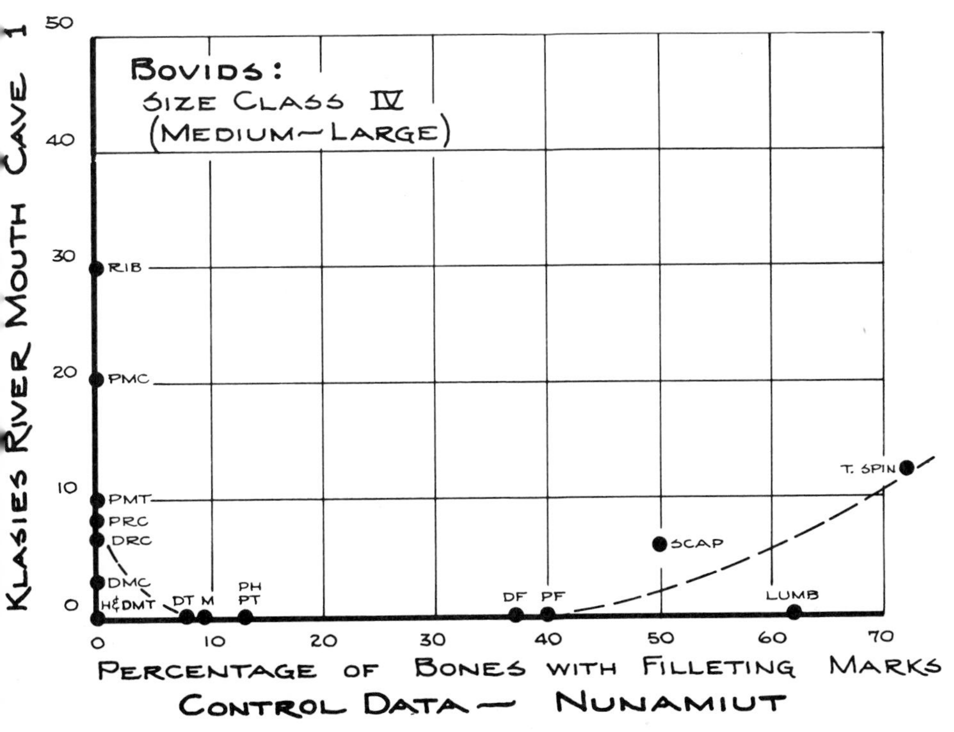

Figure 4.40 Comparison between filleting marks in a Nunamiut control population and on large bovids from Klasies River Mouth. DF, distal femur; DMC, distal metacarpal; DMT, distal metatarsal; DRC, distal radiocubitus; DT, distal tibia; H, humerus; LUMB, lumbar vertebrae; M, mandible; PF, proximal femur; PH, proximal humerus; PMC, proximal metacarpal; PMT, proximal metatarsal; PRC, proximal radiocubitus; PT, proximal tibia; RIB, rib; SCAP, scapula; T. SPIN, thoracic spines.

the skin sheath was supple. On the other hand, relatively high frequencies of chop and hack marks (Figure 4.39) on the same bones in dismemberment positions strongly suggest that, at the time of marrow cracking, the bones were in a different state than they were at the time of dismemberment. It is possible that these largely desiccated parts, scavenged from the carcasses, were actually transported to the Klasies site for processing, which included the soaking of the desiccated meat and skin in water prior to breaking open the lower limbs for marrow. Such a model of behavior would account for the baffling pattern in which chopping and hacking were commonly employed in dismemberment, whereas cutting and slicing were commonly employed in skinning the lower limbs prior to marrow cracking.

In the case of the ribs, I strongly suspect that most of the filleting marks were inflicted while removing meat from vertebrae to which broken rib sections adhered (see Figure 4.36). Checking against my notes, I observe that almost all the ribs with inflicted marks are short sections of proximal rib with filleting marks (scrape striations or short, sawing chevrons) on the dorsal surface within about 5 cm (2 inches) of the rib head. As described in the breakage data on ribs, most sections were almost certainly introduced to the site broken off, but with proximal ends still adhering to thoracic vertebrae. The marks under these conditions are almost certainly inflicted at the same time marks were inflicted on the dorsal spines of the thoracic vertebrae.

The impression one gains from the breakage and inflicted marks is that these parts were partially dried and, before filleting, had generally already been addressed by gnawing carnivores and/or carrion-feeding birds. Given these conditions, the vertebrae with attached rib ends were most likely introduced with patches of adhering and partially dried meat, which required picking off the bones or even soaking. Consumption was of small strips and strings of naturally dried meat, possibly sometimes reconstituted by soaking. Before this interpretation is secure, we need some control data on the processing for consumption of dried meat from anatomical parts such as the thoracic vertebrae.

Summary

This analysis has led to the recognition of some provocative patterning. First, there is an overall pattern in anatomical-part frequencies, frequency of gnaw marks, patterns of inflicted dismemberment and filleting marks, as well as contrastive patterns in breakage of bones—all of which tell the same story: small animals were selected, transported, and processed as essentially fresh carcasses not previously ravaged by carnivores. These small animals were introduced into the site in frequencies that betray a bias favoring meat-yielding parts (see Figure 3.18). All the data are consistent with the inference that the hominids were killing the small animals for meat.

In marked contrast are the large-animal parts preferentially introduced to the site. The large-animal parts are frequently scarred by animal teeth and bear other evidence of nonhominid gnawing. They exhibit patterns of placement for dismemberment and of inflicted marks different than those seen for the small animals. Still further contrasts are demonstrable in the frequencies of inflicted marks and parts processed for meat, as indicated by filleting marks versus marks indicative of dry carcasses, such as chop marks.

Statistically speaking, the exploitation of the large animals was in favor of anatomical parts of exceedingly marginal utility, lower-limb bones for marrow, and horn-core sinuses for pulp, as well as the lower margins of mandibles—all parts that tend to remain generally unexploited by other carnivores at their feeding sites. Considerable processing of these marginal parts was carried out at the site for what must be viewed as very small and hard-won tidbits of food. Rarely, parts from large-animal carcasses were introduced with the seeming intent of consuming meat. This is indicated by a pattern of slicing cut marks on the upper-limb bones—which are, however, rare at the site. This sugggests a gourmet selection of meat from relatively unravaged carcasses. On the other hand, the scapula and segments of the axial skeleton were occasionally introduced, seemingly scavenged from already ravaged and dry carcasses. These parts were processed for the adhering strings of naturally dried meat, and may have even been occasionally soaked to reconstitute the adhering meat. Soaking may well account for the seemingly incompatible coincidence of both hacking and slicing marks on the lower-limb bones, the latter presumably inflicted in preparation for marrow cracking by a distinctive technique.

There seems to be little doubt that there are two different sets of exploitational tactics indicated by the animal bones at Klasies River Mouth—the hunting of small animals and the scavenging of parts from large species. Other data from the marine resources at the site further attest to the casual picking up of edible foods along the beach in front of the site. We have evidence for hunting, but it is not at a scale that most would expect for such relatively late hominids. We have abundant evidence for scavenging, yet in recent years such tactics have commonly been dismissed as unlikely among our early ancestors:

> I am not trying to say that early man never scavenged. . . . Of course, to supplement his newly acquired taste for meat, these stone-age men would have scavenged when the reward was worth it and the risks not too great. We think, however, that it is more likely that man acquired his taste for flesh, like the chimpanzee and the baboon, by hunting small creatures for himself. During the birth season calves and fawns are easy prey if the hunter can manage to outwit the mothers. (van Lawick–Goodall and van Lawick 1970:28–29; from *Innocent Killers* by Jane and Hugo van Lawick–Goodall. Copyright © 1970 by Hugo and Jane van Lawick–Goodall. Reprinted by permission of Houghton Mifflin Company.)

It certainly seems to be true that scavenging was a regular and perhaps important part of the subsistence repertoire of the Klasies River Mouth hominids. Thus the exploitation of small species and calves and fawns, as suggested by van Lawick–Goodall and van Lawick, appears as a surprisingly late form of behavior.

To gain a slightly more comprehensive view of the overall pattern of adaptation practiced by the MSA hominids, can we consider other properties of the Klasies data? If so, what are the implications of this glimpse into the past for the models many of us are fond of creating as to the character of life at the very dawn of our entrance on the evolutionary stage as hominids? To this issue I now turn my attentions.

CHAPTER **5**

Hominid Subsistence Ecology and Land Use

Subsistence Tactics

In addressing the topic of subsistence of Klasies River Mouth, some of the first things to be said are rather impressionistic in character. For instance, the overall behavior that seems indicated by the faunal remains is perhaps best characterized as rigid. The food-procurement tactics seem to have been executed with a remarkable kind of repetitive sameness. This seems to be equally true for the character of the consumer demands of the hominid group. I have comparative reference to the types of behavior that I have often witnessed among modern hunter–gatherer peoples both in the American Arctic and in Australia. I might best describe the optimal behavior I have in mind as a highly skillful adjustment of achievements to goals, as well as adjusting goals to the realities of both means available and achievements realized.

Standing behind this *flexibility* is the ability to plan differing tactics for achieving variable goals, depending upon the social demands or economic conditions of the group at the time. The tactical goals of food producers continuously change as the perceived food demands of the group change with respect to feeding security, size, age structure, and so forth. These adjustments of realities to demands, and in turn demands to realities, is perhaps nowhere better illustrated than in the area of the differential use by modern hunter–gatherers of animals of different body size. Among modern hunter–gatherers, sharing is one of the most obvious examples of the adjustment of the demand for food to the amounts of food available. If a large animal is killed, then the hunter generally returns as much of the large animal as possible to the home base, then shares the food out to a large population of recipients. The social scale of the sharing unit is at least a partial function of the size of the food package made available. For instance, if a hunter returns with only a small rabbit, it is shared perhaps among only his immediate family. Hence the scale of the sharing is adjusted to the size of the food package available (see Binford 1978:139–144, 165–166).

This sharing means that human hunter–gatherers are able to absorb, without loss of food, animals of a wide variety of sizes; these would be equally processed and consumed. The patterns of consumption would not necessarily correlate inversely with the package size, as would be the case if there was no flexibility in the potential size of the consumer unit. Put another way, a large animal would be exploited as extensively as a small animal, the only difference being the size of the consumer unit sharing the food. In contrast, if the consumer unit was inflexible and producers obtained foods in differing package sizes (and no storage options were available), we could then expect differing degrees of utilization or patterns of consumption to characterize species of different sizes. Large animals would be less extensively exploited, perhaps in terms of only the most choice parts, whereas small animals would be exploited more fully as a function of their size relative to the consumer demand of the feeding unit.

Among modern humans, in whose societies sharing is a universal option, the degree of exploitation is not generally variable among animals of different body size; only the size of the participating consumer unit varies. This means that, from the perspective of faunal analysis, the package sizes introduced to a home base or consumer location (1) may be tremendously variable in size, and (2) all sizes may be equally extensively exploited even in the absence of storage techniques. Only the rate of input to the group would be apt to condition how extensively any given animal would be exploited—that is, eaten down to the last marginal but edible morsel. If for some reason there was a rapid input of large body-sized food packages into a group, it might not be able to expand the size of the consumer unit sufficiently to absorb the glut of resources; and if given no storage options, we could expect that consumption would then tend toward a "gourmet" pattern in which marginal foods would be abandoned in favor of more choice anatomical parts.

This ability to adjust the size of the consumer unit, and hence the consumer demand for resources, to the size of the food packages that must be obtained, is just one of the overall strategy characteristics of modern hunter–gatherers.

Another flexible characteristic is that the food procurement work schedule is adjusted to the perceived consumer demand. When there is plenty of good food in storage, hunters basically do not go hunting. Similarly, if several hunters have been simultaneously successful and all shared out so the consumer unit is well provisioned, then they do not seriously go hunting for a while. (They may go out into the field to monitor game movements, but will not normally make a kill unless storage is a real possibility.) Other things equal, the hunting schedule is adjusted to the perceived demand for food.

In similar fashion, hunters tend to work with a "prey image" that is consistent with their perception of consumer demand. When they go out, they know roughly the scale or size of the consumer units that they seek to serve. This means that they know roughly the total quantity of the food needed, and they adjust their prey image according to their estimates of consumer demand. As a simple example: If I have no storage capacity and have only three people to feed, it makes no sense to go elephant hunting! On the other hand, it I have a storage potential (the capability to freeze the meat and consume it over a considerable period of time—a gain in time utility from resources), then it might well enhance my security to take elephant. To be sure, there are opportunistically taken animals that may sometimes exceed both the "search image" and the ability of the hunters to utilize the produce (see, for instance, the reconstruction by Adam [1951] of events that took place at the north German site of Lehringen, suggesting that early man exploited only parts of an elephant's trunk and skull.) Serendipity would not, however, rule for such events as planned animal drives or major cooperative kills, in which the search image would by design always be in terms of large quantities of meat. If such cooperative mass kills were involved, we could expect either (1) an effective storage potential, or (2) a really large consumer unit. Lacking these conditions, a gourmet consumption strategy would most certainly rule the use of overabundant resources that might result.

The ability to adjust work schedules, search images, and sizes of consumer units ensures that in general among hunter–gatherers, there is little differentiation in the levels of exploitation characteristic of animals of different body size. The only general condition that might influence shifts in the levels of exploitation would be the degree of disjunction between food available and consumer demand. If food gluts occurred because of obtaining very large animals or large numbers of animals at once, so that supply exceeded demand, then we could expect an adjustment to the focal exploitation of the most choice parts of the meat supply along with an abandonment of the least desirable parts.

Short-term gluts may occur just like short-term shortages. The flexibility of tactics that producers may follow permits the adjustment of means to ends and on occasion, ends to realities, ensuring that in the long run there are few "excessive" episodes where food is wasted or where consumer strategies regularly covary with package size differences.

Against this backdrop, what seems to characterize the faunal materials from Klasies River Mouth Cave 1 with regard to variations in food package size and consumer strategy? Several points seem most important.

1. Among small bovids (animals weighing between 20 and 100 pounds [9–45.4kg]), complete animals and animals with only the lower limbs removed were introduced to the site. Such introductions occurred with about 26% of the animals of the small-body-size class represented at the site. More commonly, selected segments of the upper-front limb were introduced, culled to the most productive or most gourmet choice, the scapula. On the other hand, the parts suspected as yielding even greater food (upper-rear leg) were generally not introduced as culled parts.

I have viewed this situation as most likely reflecting primary consumption at the points of procurement most of the time, since the introduction of second-order parts cannot reasonably be attributed to scavenging behind other primary predators, because no animal gnawing is evident on the bones of the small animals. Of extreme importance is the fact that lower limbs processed for marrow, mandibles broken for edible pulp, and in general the introduction for use of marginal parts did not characterize the exploitation of the small animals at Klasies River Mouth. The hominids behaved as if there was adequate food available and marginal tidbits could be ignored. Nevertheless, the edible parts of a complete cape grysbok (the most common small bovid) would be about 13 pounds (5.9 kg), and for an upper-front quarter (the most commonly introduced piece-butchered part) would be around 1½ pounds (.68 kg), and the head around one pound (.454 kg). These small package sizes certainly suggest that the consumer units were very small indeed and that the planning depth was very short; that is, they generally ate choice parts in spite of the fact that the quantities available would certainly not last very long.

This pattern does not imply much food sharing, and certainly does not suggest an attempt to use all of the small animal so as to maximize the scale of the consumer pool through sharing. The consumer units appear to be individuals most of the time, and very small groups on the rare occasions when complete and near-complete animals were introduced. Sharing may have characterized the units feeding at the point of procurement, but certainly the transport of parts back to the cave does not seem to have been carried out with the aim of maximizing the size of the consumer unit.

Given what we have seen of utilization among the small animals, we might be led to expect that if a large animal was available, there would be an extreme gourmet pattern, with the choice parts commonly returned and marginal parts completely ignored, because a large animal would provide food for a much larger consumer unit than the little Cape grysbok. On the other hand, if sharing was extensive and the consumer group large, we could expect that a very high percentage of the usable foods would be introduced. What pattern is observed?

2. The larger the animal, the more marginal the parts that were generally introduced to the site (see Figure 4.19). Metapodials and parts of the head from the larger animals were most commonly returned to the site at Klasies.

I have already presented what I consider to be an overwhelming case favoring the view that the parts of large animals were scavenged from the kill and death sites previously ravaged by nonhominid scavengers. The anatomical-part frequency data, the pattern of animal gnawing, the evidence of dismemberment when stiff, and the processing investments in very marginal foods seem to me compelling proof that MSA man was not hunting these animals: (1) the larger animals were being scavenged and (2) the package sizes that were regularly introduced to the site from large individuals were very small, perhaps even smaller than from the small animals. These consisted of metapodials, which on processing yielded only a few ounces of marrow; *Tragelaphine* horns processed for the pulp inside the horn sinuses; mandibles that were processed for small quantities of pulp below the tooth rows; and, most of the time, partially desiccated parts of the axial skeleton that yielded some strings of naturally dried meat and perhaps skin. (3) In addition to being small, these parts were also very marginal food sources.

These conditions all support the view, obtained from the small animals, that package sizes were small and introduced in anticipation of individual feeders or at best very small consumer units. Planning depth was certainly very shallow and the actions were labor intensive. I might mention that there does not appear to have been any attempt to provision the site even when meat from the large animals was available. Meat-yielding parts returned to the home site were gourmet choices. This suggests that there were no well-planned search or discovery strategies used in locating carcasses, such as regular observation of vultures or attempts to locate and exploit carcasses before they were exploited and nearly exhausted by other scavengers. The picture one gets is of a hominid taking tidbits from carcasses as he happens to encounter them, rather than making a concerted effort to find and exploit carcasses early in the sequence of scavenger attrition following a death or kill. When such high-yield carcasses were encountered, however, they were not extensively exploited. High-yield gourmet parts were occasionally returned to the site, possibly for cooking.

This view of the behavior standing behind the faunal remains at Klasies River Mouth is at considerable variance with the prevailing opinion that almost universally considers all the fauna to have resulted from a single set of tactics; for example, hunting. For instance, noting the higher frequency of larger mammals represented in terms of MNIs, Klein generalizes, "MSA hunters preyed mainly on medium-to-large ungulates and generally avoided both the larger carnivores and the largest, most dangerous herbivores (rhinoceroses and elephants)" (1977a:120). Clearly, for Klein the fauna at the site result from hominid hunting, which in his view had been filtered through the "schlepp effect" (Klein 1976:87; 1980:229–230), thus accounting for the discrepencies between anatomical-part frequencies found at the site and those occurring in a living animal. The citation of the schlepp effect implies another view of the past; namely, that Klasies River Mouth Cave 1 was a base camp.

Was Klasies River Mouth Cave 1 a Home Base?

If we accept the functional association of stone tools and animal bones occurring together as an operational definition of a home base (see Isaac 1971), then certainly Klasies River Mouth Cave 1 was a home base. On the other hand, if we take a more probing view of what home bases are and what they imply, we may be forced to different conclusions. Home bases are basic to the kinds of adaptations that we can see among most of the world's hunting and gathering peoples known from recent times. As Glynn Isaac (1978) has correctly pointed out numerous times, home bases imply almost all the essential features of the relatively unique set of behavioral–organizational features characteristic of modern man's way of dealing with his environment. Most fundamentally, home bases imply provisioning tactics. That is, producers move out into the habitat seeking and obtaining foods that are then transported back to a central place as a contribution to the provisioning of the group, or at least of the individuals living there. There seem to be two major components, organizationally speaking, to the idea of a home base: (1) that a group lives somewhere, and (2) that this group is provisioned by virtue of the actions of group members who accumulate foods primarily consumed at the living place.

A provisioning model of human subsistence assumes sharing, and sharing generally assumes that producers seek to obtain foods in package sizes that exceed their individual food demands; otherwise there would be nothing to share. The model of land use that is demonstrably appropriate to the home base or central-based foraging model of subsistence is one in which the basic life space of a group is within a site—the home base—and producers forage out of this life space into the surrounding environmental space in search of foods and other necessary provisions (firewood, water, etc.). The points of procurement are special-purposes locations or points of exploitation by the foraging producers. Provisions obtained at such locations are then returned to the home base, and sharing normally follows. The scale of sharing varies with the amount of provisions obtained and the conventions of the group as to what is considered appropriate to share.

This model of land use is illustrated in Figure 5.1, in which the home bases are considered the basic life spaces of the group, and the home base with its resident group is provisioned by producers foraging out into the environment and obtaining needed goods at special purpose locations, or points where provisions are obtained. These provisions are then returned to the home base, where they are processed and consumed. Home bases as basic life spaces are the places where sleeping occurs, care of the young and aged (protected life space) is carried out, reproductive acts most often occur, and so forth.

My earlier studies of modern hunter–gatherers (see Binford 1978, 1982b) have shown that home bases are residential hubs of hunting and gathering systems, and because of this central or focal role they can be expected to be quite variable in their content from one occupational episode to the next. First, the variability derives from the organizational complexity of a home base because so many of the basic life functions are centered in such places. Food consumption, sleeping, social life, reproduction, and care of the young are all localized in such places. Second, there is a temporal or sequential source of variability, which derives from the details of the history of the occupation. All the food procurement, processing, and transport tactics practiced by a group vary with the mixes of success and failure experienced by the group while living at the site.

Each separate occupation generally differs one from another, in that, although the same basic repertoire of tactics may have been available for use by the occupants, the particular mix of successes and failures and, in turn, the mix of primary, secondary, and tertiary tactics triggered by differences in failure rates, varies with the situational conditions of the group during the terms of the occupation. Such responsive behavioral variability to perceived conditions by the ancient actors results in a wide range of content variations in the archaeological remains at home-base sites.

There is a further complication contributing to site content variability that derives from long-term spatial mobility (see Binford 1983a:379–386): the way a place is used is relative to the placement of the home base. As home bases are moved, the economic geography of the area changes relative

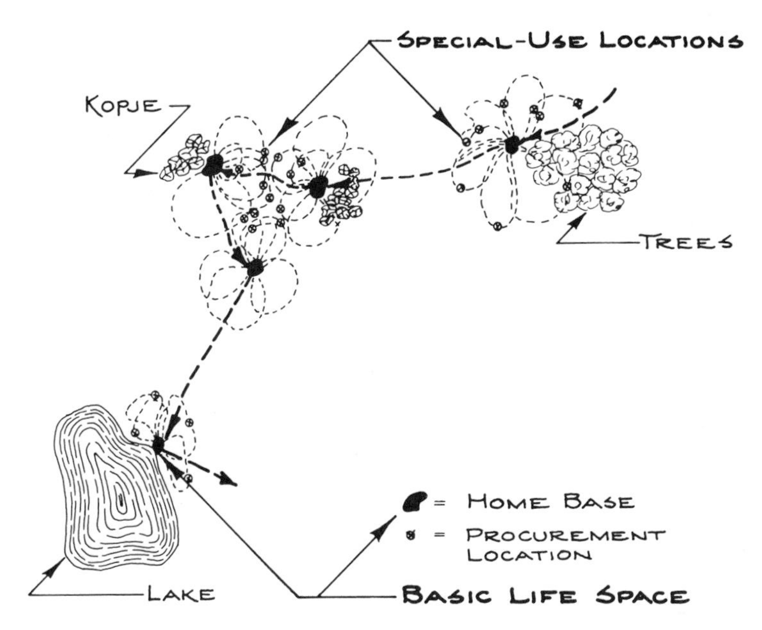

Figure 5.1 Model of home-base or central-place land use.

to the home base, and both former home bases and other more specialized use locations may change in the ways they are regularly used (see Binford 1982b:18–20). The positioning of the system in space ensures that the content of sites, particularly stratified deposits, will appear variable and exhibit patterns of structural–formal differentiation and content variability between occupational episodes. Thus we may argue that, given sharing as a basic component of home-base living, we should see (1) little differentiation in the scales of exploitation evidenced by the remains of large and small animals. We could expect some accommodation in terms of transport and alternative field processing of food packages of vastly different size, but little variation in the degrees of exploitation, since the flexibility afforded by sharing ensures that the size of the consumer unit is expandable through sharing to accommodate large food packages that might be introduced on a regular basis. If this increase in the sharing scale and hence consumer pool is not regularly possible, then modern hunters adjust their search images to food packages appropriate to the consumer-demand units. We should not see major differences in consumer strategy regularly associated with animals of different body size unless there is a storage potential. In like fashion, if we have home bases we should expect (2) substantial variability in the archaeological content of home-base sites, deriving from the situational differences experienced by the group while carrying out their basic strategies for procuring food processing and returning it to the home base for consumption. As Yellen (1977:135) has noted, these sites should be variable even if only in response to differing lengths of occupational episodes.

As already pointed out, the data on processing different animals of different body size seem clearly indicative of (1) differing procurement tactics for animals of different size, and (2) biased selection of parts from the large animals that differed greatly from the parts selected from the small-animal carcasses. The small animals were butchered, processed, and transported with respect to considerations of the distribution of fresh meat on their skeletons. This may well have been true of the very young individuals from otherwise large species, such as Cape buffalo and *Pelorovis,* although the bones from the very young individuals from these species have not been specifically analyzed with this question in mind. In marked contrast, the bones from the large animals were selected and processed primarily in terms of limited amounts of bone marrow, marginal foods recoverable from the head, and possibly small amounts of naturally dried meat remaining adhering to parts of the axial skeleton after other scavengers had exploited carcasses.

These conditions imply two very important things about the behavior of the hominids responsible: (1) they appear to have taken live animals only rarely, and (2) these were small animals that they could overpower on discovery. In short, there is nothing in the data from Klasies River Mouth to suggest technologically aided hunting, or even tactical hunting as such. The taking of these small bovids and the young of larger animals could have been done quite opportunistically as the hominids happened to encounter them while feeding through brushy cover during the day. The data on the biased introduction of front-leg parts from the small animals is strong evidence that most consumption took place out in the field, and only parts left over were regularly returned to the site at Klasies River Mouth. Much primary consumption seems to have occurred at points of procurement.

Such an interpretation is also consistent with the character of the parts introduced to the site from the larger animals. In almost all cases, the introduced parts were subjected to substantial processing prior to consumption. In many cases, the lower limbs were dismembered while stiff and desiccated, yet appear to have been skinned while soft and supple. I have suggested that this could only take place if the lower limbs were removed from partially desiccated skeletons and then carried to a water source where they were soaked prior to being processed for marrow. This same condition, reconstituting through soaking, might well stand behind the heavy-handed hack marks on the parts of the axial skeleton from large animals, which are also commonly carnivore-gnawed. These parts could also have been reconstituted by soaking, so that remnant meat could be cut from the bones that had previously been ravaged and were dry and stiff.

Finally, there is evidence of cooking at Klasies River Mouth, and this evidence is most provocative in that it is internally consistent with the biased preparation of the rear leg, a prime meat-yielding part that can be butchered so to remain encased in its own skin. This choice method of food preparation, which preserved the natural juices of the meat, was almost exclusively conducted on parts of the rear leg—a part that as we have seen, was rarely introduced in its meat-yielding form to the site. Nevertheless, it received most of the processing attention; head parts were second. Anatomically speaking, the meat-yielding parts most commonly introduced, such as the scapula and upper-front leg, were not generally cooked. This total pattern is consistent with a past behavior in which hominids returned food elements to the site at Klasies River Mouth, not necessarily as provisions for sharing, but as parts intended for processing. A pattern of primary consumption–feeding at the points of procurement, with the occasional return of parts requiring considerable processing to safe locations, accommodates the facts at Klasies River Mouth much better than does a picture of hominids living in home bases and provisioning such places by tactical hunting of large animals. A processing focus seems more consistent with the facts than a provisioning focus at the site. There are still other reasons for home-base skepticism.

I think it is likely that the subsistence activities standing behind the introductions of animal foods to the site can be summarized as:

1. The occasional killing and transport of meat-yielding parts into the site from small bovids and the young of larger bovids.
2. The scavenging from carcasses of larger bovids, parts of usable but generally marginal utility. These are mainly marrow-yielding lower-limb bones; head parts, including the horns of tragelaphines, which were processed for the pulpy contents of horn sinuses; and dried-out remnants of the axial skeleton. All these parts seem to have been processed at the site prior to consumption.
3. Opportunistically collected foods recovered from the upper storm-beaches near the cave. Both collected (shells) and scavenged foods (seals) were introduced to the site.

The Ecology of Scavenging

This generalized view of subsistence of Klasies demands a consideration of the activity patterns of the Klasies hominids regarding the ways the two different exploitation tactics, hunting and scavenging, were organized and differentially executed. In order to address this problem I have sought clues in the character of the species that appear to have been hunted versus those that were regularly scavenged.

All the animals that seem to have been killed or at least obtained for their meat are moderate to small in size, generally nocturnal in their feeding activities, and territorial and solitary in their social behavior. In addition, they have preferences for scrub-brush types of cover and habitat. On the other hand, the larger animals regularly scavenged by Klasies hominids seem to have in common a different type of behavioral repertoire. They tend to be nocturnal drinkers. I have previously described the dynamics of an African water source in a very dry Kalaharian environmental zone (Binford 1983b:62–70). It was noted that carnivores (hyaenas and lions) tended to drink after dark and in the early hours of the morning. In addition, they tended to focus their behavior on the waterholes beginning around sunset and intermittently returning during the night. My observations on hunting by these carnivores suggested that most kills of the animal species that avoided the waterholes occurred away from the waterholes at night. What was missing from my experiences was the larger-body-size, more social bovids, which generally inhabited somewhat more moist settings. Fortunately, there is a fine study by Ayeni of the patterns of drinking by African species at waterholes in Kenya's Tsavo National Park.

> A basic pattern of waterhole utilization dominated by small (adult-size) species during day-time 0600–1800 hours and larger species at night 1800–0600 hours is described. The separation in times of arrival and departure peaks of waterhole utilization, and average coincidence of percentages of paired species populations are used to show that big-game attained a measure of time-spaced ecological separation at the waterholes. (Ayeni 1975:305)

Ayeni found not only that species varied seasonally (dry versus wet seasons) in their visits to waterholes, but that they varied in a most significant way in the degree to which they visited waterholes coincidentally with carnivores. Carnivores were, as noted earlier, mainly visitors to the water sources during darkness. Not surprisingly, Ayeni found that the frequency of freshly killed animals (prey) recorded in the immediate vicinity of the waterholes was a general function of the coincidence of the prey animals drinking at the same times as carnivores, excepting of course the differential ability of the carnivores to take animals of very large size; for example, elephants.

Table 5.1 summarizes Ayeni's data on density of game in the region, numbers of animals of different species recorded as prey around waterholes, and the percentages of total drinking visits to a waterhole that were coincidental between carnivores and the species considered. For instance, if all

TABLE 5.1

Coincidental Appearance of Animals at Waterholes with Carnivores, and Their Relative Death Rates[a]

	Density of species per km of road, 1973–1974 (1)	*Percentage coincidence with carnivores at waterholes (2)*	*Column 1 × Column 2 (3)*	*Numbers of individuals killed near waterholes (4)*
Zebra	1.127	.10	.11	6
Elephant	1.103	.40	.44	0
Impala	0.533	.07	.04	0
Waterbuck	0.513	.12	.06	11
Kongoni	0.480	.09	.04	2
Oryx	0.453	.01	0	4
Buffalo	0.33	.38	.13	19
Peters	—	—	—	—
Gazelle	0.303	.07	.02	0
Warthog	0.197	.03	.01	0
Ostrich	0.080	0	0	0
Giraffe	0.070	.24	.02	1
Eland	?	.24	(.13?)	14
Rhinoceros	?	.39	?	0

[a] Data from Ayeni (1975:Tables 1 and 5).

drinking visits of a species, such as warthog, were during hours when no carnivores were observed to drink, then the percentage coincidence would be 0%. On the other hand, if 40 out of 100 drinking episodes recorded for eland were coincidental with hours also recorded for carnivores drinking at the waterhole, the percentage coincidence would be 40%.

Inspection of Table 5.1 demonstrates that the frequency of animals occurring as prey around the water sources is a combined function of the number of animals coming to the waterhole and the levels of coincidence for these visits with carnivores. Figure 5.2 illustrates this relationship nicely. Except for the elephant, which is generally considered to be exempt from heavy predation due to its size, the frequency of observed kills is a linear function of density and coincidence at the waterholes. This means that for a scavenger exploiting animal remains around a waterhole, the carcasses available to it will normally be a function of the numbers of potential prey and their drinking schedules relative to the drinking and activity schedules of the predators.

It should be clear that although impala, kongoni, oryx, and zebra are all much more common in the environment, they are rare in the population

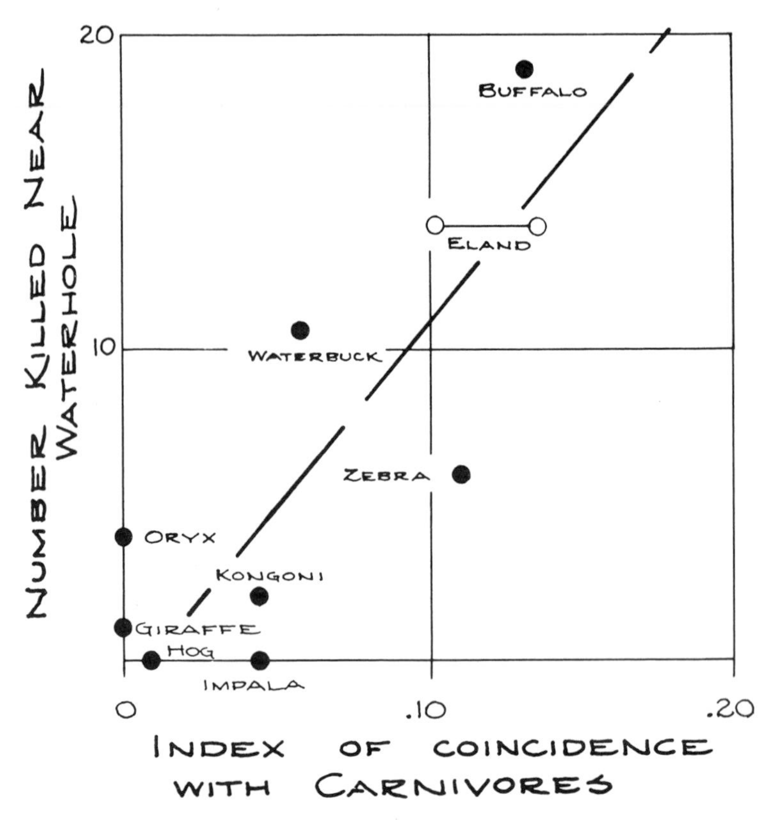

Figure 5.2 Correlation between frequency of species as prey and their coincidence with carnivores at waterholes.

of animals killed by predators around the waterholes. On the other hand, eland and buffalo are far less common in the environment, but because their activity schedules overlap that of the carnivores, particularly in terms of drinking schedules, they are therefore more commonly killed around the waterhole. This means that in the relative frequencies of species available to a scavenger around waterholes, one is getting a view of the population of animals that tend to be nocturnal in both their feeding and drinking schedules, as well as of the numerically common and relatively large-body-size diurnal feeders who overlap carnivores in the morning and evening hours. The species that tend to be exclusive diurnal feeders and drinkers should only rarely appear in the population of carcasses killed by predators around a waterhole. As Ayeni points out, the species that tend to utilize the waterholes during the day are more apt to be those that dissipate heat by panting rather than those that sweat. These species tend to have generally smaller body sizes and to be the species most adapted to very dry environmental conditions. Thus, in spite of their presence in the habitat, they would not show up as major components of animals killed by nocturnal predators around waterholes. This has major implications for the type of direct environmental interpretations Klein (1976:79–80; 1980:240–241) tends of make from species frequencies.

Given this knowledge about the population of animals apt to be available to a scavenger around an African waterhole, we can go back and examine the species composition of the animals that appear to have been scavenged by the occupants at Klasies River Mouth Cave 1. These species were the giant buffalo (*Pelorovis antiquis*), the Cape buffalo (*Syncerus caffer*), the eland (*Taurotragus oryx*), the blue antelope (*Hippotragus leucophaeus*), the kudu (*Tragelaphus strepsiceros*), and the bushbuck (*Tragelaphus scriptus*). In all cases in which we know the behavior of these animals, they tend to be morning and evening drinkers, or nighttime drinkers, and most are basically nocturnal feeders. In short, these are the species we could expect to be differentially killed adjacent to waterholes by African nocturnal predators. This fits well with Klein's repeated citation of the fact that the age profile of Cape buffalo killed by lions in the Serengeti National Park is essentially the same as that observed at Klasies River Mouth. The pattern of waterhole prey also explains the general lack of bushpig or warthog at Klasies River Mouth, which has so puzzled Klein (1975b, 1976). Ayeni's data (Table 5.1) shows that the pigs are midday drinkers; hence these species rarely appear as predator meals around waterholes. Klein has repeatedly suggested that the lack of pigs at Klasies was because MSA hominids avoided hunting them because they are dangerous animals (Klein 1976:83).

On this point Klein is almost certainly correct, but the hominids also seemingly avoided hunting all animals whose size exceeded about 90 pounds (40.8 kg). The scheduling explanation for pig absence at Klasies is further substantiated by the bones of animals falling into size class III. In this group are both species that visit waterholes primarily at dusk and dawn (kudu and *Hippotragus*), as well as essentially midday visitors such as gemsbok, wildebeest, hartebeest, and bastard hartebeest. It is interesting that the anatomical-part frequencies for the diurnal drinkers (bastard hartebeest, wildebeest, and hartebeest) tend toward the pattern of meat exploitation described for the smaller animals. The only difference is that upper-rear legs are somewhat more regularly represented than is the case for the small animals. On the other hand, the kudu, bushbuck, and blue antelope are all more commonly represented by the head and lower leg pattern we have noted for the larger animals. These we know to be drinkers whose schedules overlap that of the carnivores.

As a demonstration of this difference between the midday drinkers and the dawn–dusk drinkers, I have prepared the graph shown in Figure 5.3. Here we see the percentage of unfused bones for a series of anatomical parts arranged from left to right, so that parts on the left are the parts which in other bovids tend to fuse earliest in the maturation sequence. If all the animals had been introduced as complete skeletons and there was a single age distribution to the population introduced, then we should see a relatively smooth curve rising from left to right across the graph with the last two entries being roughly equal. This is nearly the curve realized for the nocturnal or morning and evening drinkers, the tragelaphines clearly suggesting that there was a unimodal age distribution, not biased in favor of young animals, evidenced in the bones transported to the site by the hominids. They had selected bones in roughly a random fashion with respect to the ages of the animals from which the bones were culled. On the other hand, bones from the midday drinkers, the alcelaphines, exhibit a very different profile. There is a high frequency of unfused bones from the pelvis as well as the proximal femur and distal femur, while there is an equally, extraordinarily low percentage of unfused bones of the lower-limb bones (distal tibia, distal metatarsal, and proximal tibia). Here we see the exploitational pattern split between meat-yielding parts introduced from very young individuals, and the non–meat-yielding bones of the lower legs—heavily biased in favor of adult animals. Scavenging of the tragelaphines is consistent with the view that they were prey of lions and hyaena taken mostly at night around waterholes, whereas the capture or killing of very young alcelaphines by the hominids would have been accomplished primarily during the midday hours, when they tend to drink and to coincide with one another at water sources.

Several points are clear: (1) The hominids were essentially diurnal species. All the primates are essentially diurnal and we continue to be so. Our eyes certainly limit the types of activities what we might engage in at night. I think there is little doubt that the early himinids were only active during the daytime. (2) The hominids scavenged primarily from the remains of the prey taken near water sources by essentially nocturnal killers. This could be done by hominids with no undue exposure to predation if they visited water sources during the middle of the day, as do the daytime drinkers. The data from Klasies River Mouth Cave 1 strongly suggest that the hominids regularly used a site near a water source also generally used by other animals. There they scavenged large animal parts, which were later introduced to the Klasies living site.

To illustrate the likely schedule of a generalized hominid relative to other animals he is apt to have interacted with in an African setting, I have superimposed the feeding schedules actually recorded for the modern gorilla

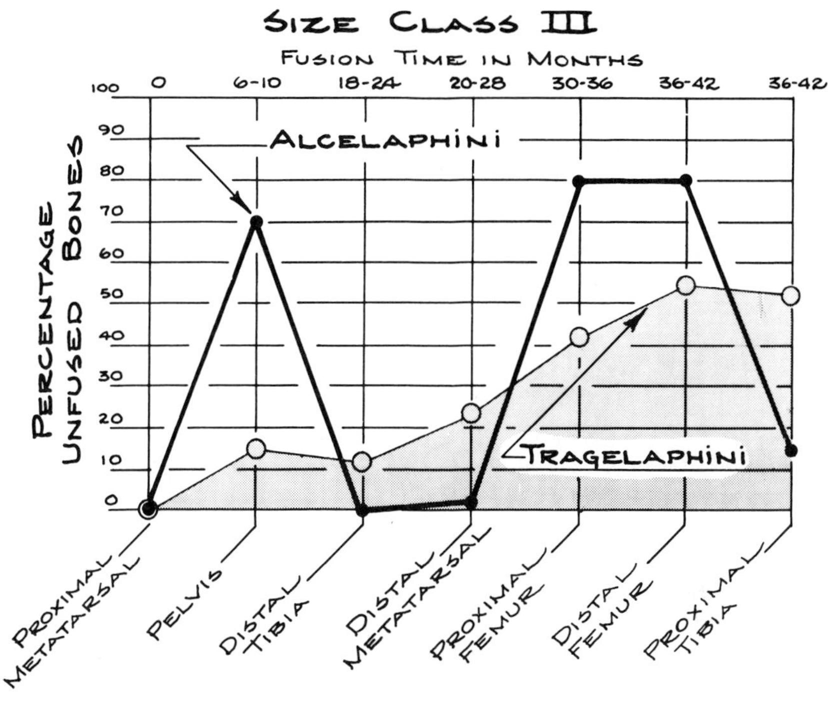

Figure 5.3 Comparison of unfused bone percentages between night and day drinkers, size class III, from Klasies River Mouth.

over the data on waterhole visits collected by Ayeni (1975) for a variety of African species. The pattern in Figure 5.4 suggests a general pattern and a niche for the early hominids. A likely scenario might be as follows: The hominids awake at sunrise and begin feeding in the early morning. This feeding is likely to have been in scrub-brush habitats away from the normal water sources used by the typical African predators. Feeding might generally begin to diminish after around 10 o'clock and the hominids approach the major water source, where they rest during midday in the company of other African forms that tend to be midday drinkers. During their stay at the waterhole, hominids scavenge edible parts from the carcasses that the nocturnal predators have killed in the vicinity of the waterhole. They might on occasion capture the very young of some of the midday drinkers (*alcelaphines*) at the waterhole. Around 1 o'clock, the hominids begin moving off the waterhole and again feed through the woody-brush zones where plant foods are available and where fawns as well as the small nocturnal antelope might be encountered. Finally, in the late afternoon the hominids

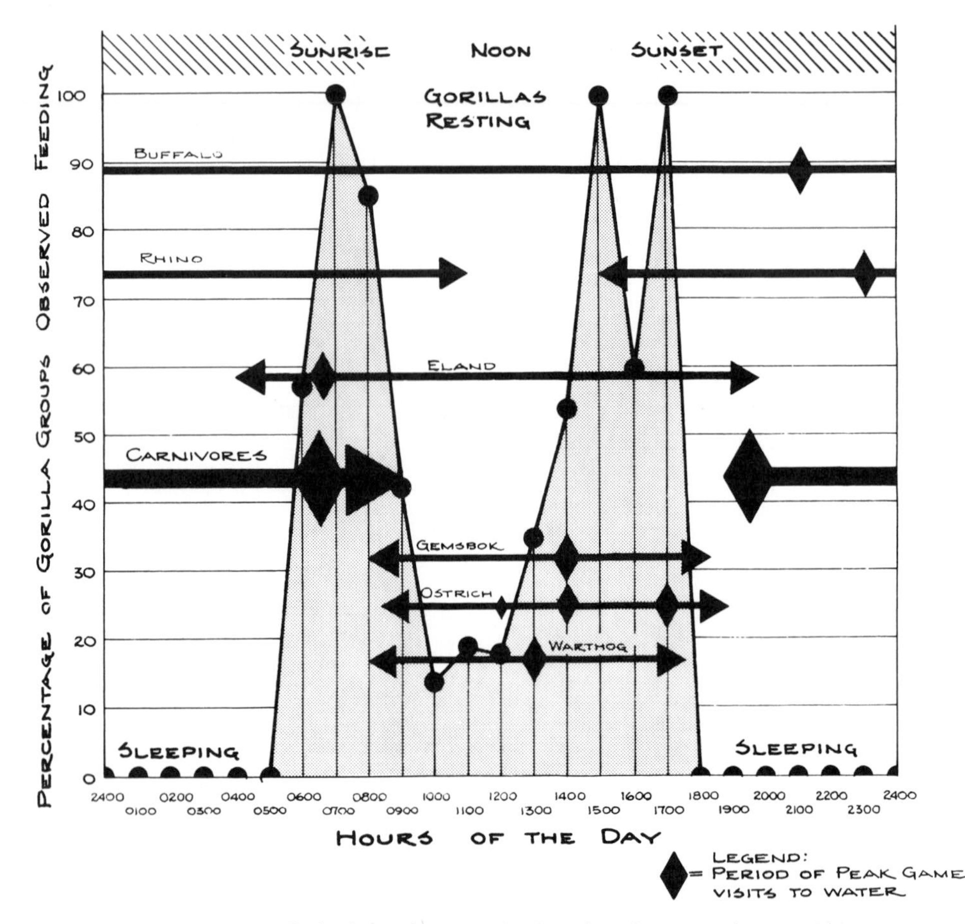

Figure 5.4 Gorilla feeding relative to drinking by other animals at an African waterhole. Gorilla data from Schaller (1963:147); waterhole data from Ayeni (1975:319).

return to a secure sleeping place, where at sunset they begin their nighttime sleep.

I suggest that the early sleeping places would never be at the watering places of the largely nocturnal carnivore and ungulate species, but would be in relatively protected areas away from the water sources. This picture suggests two major types of special-purpose location: the midday waterhole rest area, and the nighttime protected sleep area. Feeding from terrestrial sources would be carried out primarily in the scrub and brush zones between these special-purpose locations.

The objection may well be raised that the Tzitzikamma Coast is not the Tsavo game park of Kenya, nor is it the dry environments that I observed in the Nossob River area of the Kalahari. Given that it is such a different environment, how can we build a behavioral model for the species in a *fynbos* biome based on observations of animals behavior from such differing environments?

I am not saying that the ecology of the regions are alike—clearly they are different—but I am suggesting that some of the basic niches, particularly those of the predators, are similar. Certainly, too, the requirements of many of the species, which are essentially the same, cannot be expected to change from one environment to another; only their strategies of meeting their needs might be expected to vary. Perhaps the fact most fundamental to the argument here is that while lions may have relatively large territories they are essentially territorial animals.

> Each pride confines itself to a definite area in which its members spend several years, or in the case of some lionesses, their whole life. The main requisites for the existence of a pride area are *a water source and sufficient prey throughout the year, conditions existing in the woodlands and along their edges* but, for the most part, not on the plains. (Schaller 1972b:56; emphasis added; © 1972 by the University of Chicago)
>
> Territories vary in size, but do not seem to change in size much with time. Where there are ample numbers of resident prey animals, a pride usually occupies between ten and forty square miles. (Bertram 1978:105; Brian Bertram, *Pride of Lions*. Copyright © Brian Bertram 1978. Reprinted with the permission of Charles Scribner's Sons.)

In the Serengeti Plain, where some of the classic lion studies have been conducted, a very interesting fact was noted: the population of lions does not vary directly with the populations of grassland-feeding animals, such as wildebeest and zebra, but tends to be limited by the density of resident game largely made of scrub–woodland-loving species. The lions do not migrate with the herds of grass feeders during the wet season. True, they may alter their patterns of territorial use a bit, but, in general, during the wet season when the grass feeders are on the plains, the lions feed primarily on the other animals that remain in the forest–scrub margins where the lions also reside. This means that there is a major seasonality to the lion prey—resident bush-loving species during the wet season and a shift to the migratory visitors from the grasslands during the dry season. The latter, of course, come into the scrub where water and some forage is available during the dry season. Although some of the species would be different, and certainly the overall organization of the ecosystem along the Tzitzikamma Coast would certainly be different, the tempo of seasonal variability and the relationship between a resident predator (the lion) and resident prey (the bush-loving species such as the Cape buffalo, kudu, blue antelope, bushbuck, and bushpig) was probably not vastly different from the predator–prey relationship in many other environments.

Lions are essentially nocturnal feeders, as already noted. They are also not very successful pursuit hunters, being much more successful at ambush and stalking prey in cover, particularly tall grass, which incidentally is the favorite feeding place of the Cape buffalo. Both the Klasies river and the Tzitzikamma rivers would have supplied (1) watering sources for both lions and prey, and (2) tall grass providing forage for such species as the Cape buffalo and ambush cover for lions.

We do not know how the hydrology of the two rivers varied during the Late Pleistocene, but the hippopotamus is represented at Klasies Cave 1 continuously through the MSA I and II levels, and is also present in the Shelter 1A levels about 17 (the Howieson's Poort and MSA II; see Klein 1976:77–78). This almost certainly means that the rivers never completely dried up, and that there were at least substantial dry-season pools in the riverbeds throughout the Late Pleistocene, represented by deposits at Klasies River Mouth (the low-water Level 13 and, interestingly, the MSA II levels in Shelter 1A are perhaps exceptions). In any event, the periods of major accumulation were contemporary with year-round water sources in the riverbeds, providing lions and prey with reliable water sources. Such reliable water would have also been attractive to seasonal migrants, particularly during dry seasons. We can expect that under those conditions the prey of lions would shift with variations in the regional water budget, so that during the dry season, migrants taking advantage of reliable water would be targeted as lion prey. On the other hand, during the remainder of the year the resident animals would supply the local lion pride with food. In both cases, the customary drinking places along the Klasies and Tzitzikamma rivers would most certainly have been the focus of lion predation. It is quite likely that during the Late Pleistocene sequence the most common migrant into the area was the eland. According to information from Dr. J. C. Hillman,

> The eland . . . give birth to their calves over a period of approximately five months beginning in August, with peak calving around the short rains in November. At that time the cows are usually out on the plains . . . during the rains, when the animals are concentrated on the grasslands, the nursery groups can get very large. . . . As the plains begin to dry up in January and February, however, the large groups begin to break up, the eland move into the river gorges and feed on more dispersed food items. (Moss 1975:183–184)

Most observers have commented on the fact that female eland are much more mobile than males and that they tend to congregate during the peak rains on the open grassland, where they give birth and may be seen in large nursery herds (see Kingdon 1982:127–141). At the same time males tend to reside more in river gorges and maintain a more solitary territorial existence. As the dry season approaches, the congregations of eland break up into smaller and smaller units, dispersing into the woodland and river gorges where they feed on a greater amount of browse. This behavioral pattern is consistent with the seasonal appearance of eland along the coast in the better-watered *fynbos* biome, which supplies a reliable source of water during the dry season on the grasslands to the north, or to the interior of the Cape Folded Mountains. This pattern implies that during the Late Pleistocene eland young were not born on the Tzitzikamma Coast; considering body size, the year's calves that might enter the region during the grassland's dry season would by then be too large for the hominids to take, given the body-size ranges for which they seem to have been successful at hunting. We might expect that occasionally the zebra, wildebeest, and perhaps the bastard hartebeest might appear south of the mountains as dry-season migrants.

It is my guess that the fluctuations in the frequency of grass-loving species in the Cape coastal region (at least in the context of the environments represented at Klasies River) are not necessarily an expression of the expansion of grasslands into the Cape geographic zone, but instead the expansion of a more lush grassland into what is today the Karoo. Grass feeders then more commonly took to the higher biomass Cape zones during the interior dry seasons and more commonly showed up on the Cape as seasonal visitors.

In any event, the basic aspects of waterhole dynamics that I have noted for the Kalaharian zone and has been reported in the Tsavo area can be expected to have characterized the Cape zone, given the presence of lions and prey. It seems quite likely that this focused concentration of lion-killed carcasses at focal drinking spots along the Klasies and Tzitzikamma rivers were the primary points for hominid scavenging as manifest in the faunal remains at Klasies Cave 1.

Scavenging and Age Profiles

As in the case of Klein's other interpretations, he assumed a behavior—hunting—and then sought to interpret the data from the sites so as to be compatible with that assumption. In the case of his age-profile data, he has assumed hunting and then made an additional assumption regarding ages at death of the animals represented in the sites: "*There is no reason to suppose that either human behavior or differential durability seriously biased any age distribution presented here*" (Klein 1978c:203–204; emphasis added).

These two assumptions—that all the bones were in sites by virtue of the actions of hunting hominids, and that hominid behavior did not bias the faunal population so that the bones in the site actually represented the frequencies of different ages and species killed by the hominids—have permitted Klein to use the bone assemblage for direct inferences about the character of hominid hunting tactics. For instance:

> The catastrophic mortality profile that characterizes eland (from Klasies river mouth Cave 1) . . . almost certainly reflects the stone age discovery that this species is relatively easy to drive over cliffs or into other traps (Klein 1978). Similarly, the catastrophic mortality profiles that characterize some relatively small antelopes in the same deposits . . . probably reflects human awareness that individuals of all ages may be readily caught in snares laid along their habitual runs through the bush (Klein 1981a). (Klein 1982:153)

For the first time, Klein (1982) seriously considers the possibility that the hominids were scavenging; he proceeds to offer suggestions how one might distinguish hunting from scavenging. He advocates the comparison of mortality-age profiles from archaeological sites with those from "non-archaeological sites" (Klein 1982:154), presumably as a control on the consequences of human behavior as opposed to "nature" in conditioning mortality curves. Klein then points out how both "generic" forms of age profile could derive from hunting, with catastrophic curves resulting from animal drives or from indiscriminate trapping, while attritional curves might arise from stalking and more individual hunting techniques—because "prehistoric peoples were usually unable to obtain prime-age adults of a species" (Klein 1982:154). In spite of Klein's claim that both curves could result from hunting, he proposes that "the proportionate representation of different age classes in a species sample from a site provides a means of determining whether the species was primarily scavenged or hunted by the people" (Klein 1982:1515).

In demonstration of this suggestion, Klein compares the age profiles of the giant extinct buffalo (*Pelorovis*) from the famous site of Elandsfontein (see Figure 2.1 for location) with that described for *Pelorovis* from Klasies River Mouth Cave 1. Figure 5.5 summarizes Klein's data from these two sites expressed in terms of percentages of life span and percentages of the total number of individuals assignable to each age class. Superimposed over both is the idealized model of how a "normal" attritional age (mortality) profile should appear according to Klein (see Klein 1982:Figure 1). Several points should be clarified by this comparison: (1) young individuals are underrepresented relative to the model at Elandsfontein, whereas (2) young individuals are overrepresented at Klasies River Mouth. On the other hand, (3) very old individuals are generally underrepresented at Klasies. Noting these facts, Klein offers the following interpretation: "The reason that very young buffalo are proportionately much less common at Elandsfontein is probably because they were selectively removed from the record before burial, primarily by carnivore feeding" (Klein 1982:156).

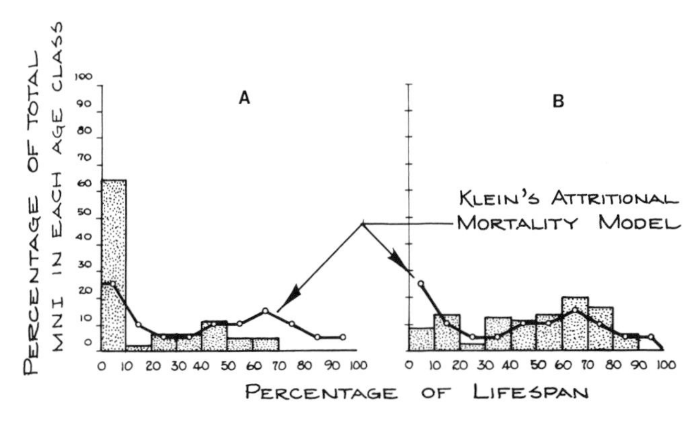

Figure 5.5 Comparison between giant buffalo (*Pelorovis*) age (mortality) profiles from (A) Klasies River Mouth and (B) Elandsfontein populations.

Although I might in general agree with this suggestion, it should be pointed out that there is clear evidence in the form of isolated tools in association with recognizable death sites (see Binford 1983b, Fig. 30) at Elandsfontein that we must also include the possibility that hominids might well have played some role in biasing the Elandsfontein data against very young individuals. Klein, however, ignores this possibility, then goes on to note that

> the much greater proportion of very young buffalo at Klasies could result from scavenging *only if* the Klasies people could locate the carcasses of the animals *before* they were located by potential competitors, particularly hyenas and lions. Lacking the relevant special senses of these competitors, it is highly unlikely the Klasies people could locate carcasses first, and from this it may be inferred that the very high proportion of very young buffalo at Klasies must reflect active hunting. (Klein 1982:156–157)

Klein's reasoning here is faulty. The biased exploitation of young and small buffalo does not require that the hominids find young or small buffalo *before* competitors on a regular basis; only that when they do encounter usable parts of buffalo, they will remove heads of small (young) animals preferentially. That a hominid scavenger would behave in this manner is perhaps best illustrated in Figure 5.6. The drawing was based on a wonderful photograph published in Lorna Marshall's provocative book, *The !Kung of Nyae Nyae* (1976:90). It shows a Bushman woman carrying the head of a male wildebeest, prior to roasting it in an earth oven. It should be recalled that a wildebeest weighs approximately 400 lbs (181 kg). Now, if we keep in mind that the extinct buffalo *Pelorovis* is thought to have been 10 times bigger than the wildebeest shown in Figure 5.6, with horns that exceed 6

Figure 5.6 Nyae Nyae Bushman woman carrying the head of a wildebeest (*Connochaetes*).

feet (2 m) across, one can only marvel that any adult *Pelorovis* skulls are represented in the site!

I think it should be clear that the bias is not necessarily in killing, but in the size of animal heads elected for transport back to the site at Klasies River Mouth. This is a situation in which the schlepp effect that Klein is fond of citing was most certainly biasing the age profiles against the larger animals. As further support for this interpretation, the reader is referred to Figure 4.4, in which it was shown that there was an increasing bias against heads as the body size increased among species. This relationship to body sizes is nowhere better demonstrated, however, than in Klein's own data on the frequency of unworn deciduous fourth premolars from the three large ungulate species, as shown in Table 5.2.

The relationship between increasing body size and the transport of increasingly younger animal heads seems clear. Klein's assumption that there is no behavioral bias in his data is clearly not justified. I think it is obvious that the age-biased graph in Figure 5.5 could easily arise if a hominid was transporting in a biased manner heads of smaller individuals. How the hominid obtained these heads is not implied by such age bias.

Given that we have already demonstrated the operation of two separate faunal procurement strategies—killing small animals and scavenging carcasses of large animals—it seems quite likely that the single age profile that

TABLE 5.2

Frequency of Worn and Unworn Milk Premolars from Klasies River Mouth and Nelson Bay[a]

		Klasies River Mouth			Nelson Bay Cave		
	Body size (lbs) (1)	*Unworn dp4 (2)*	*Total dp4 (3)*	*% Unworn (4)*	*Unworn dp4 (5)*	*Total dp4 (6)*	*% Unworn (7)*
Blue antelope	300	8	18	44.0	6	16	38.0
Cape buffalo	2000	18	32	56.0	10	17	59.0
Giant buffalo	4000	36	44	82.0	4	6	67.0

[a] After Klein (1978: Table 7).

Klein seeks to interpret in terms of hunting tactics is, in fact, the combined result of the operation of the two separate sets of tactics: killing the young calves and fawns, and scavenging the carcasses of adult animals of the same species. If so, the collapsing of age data into species categories and seeking to interpret the summary age profile, as if it referred to the accumulated product of a single set of hunting tactics, is certain to be misleading. This problem is further exacerbated by the bias in transported heads, which is fairly clearly related to size. Given these problems, I think that Klein's interpretations of hunting tactics and methods based on age profiles can be dismissed on methodological grounds.

The Ecology of Hunting

What can we suggest about the hunting strategies of food procurement characteristic of the Klasies hominids? Any predator operates in terms of the general body-size range in which it can reasonably kill, and a prey-speed range within which it can effectively overtake prey. Hominids are no match in speed for most of the bovids, so the speed range within which hominids can be expected to operate most effectively is very slow indeed, even when game is at rest. This is even true when using tools, since stalking is man's attempt to get close enough to wound the animal before it is frightened into flight. Most modern hunters are essentially "killing at a distance" while the prey is stationary. We are just not very fleet pursuit hunters. This means that the game that hominids would have been most skillful at taking were essentially stationary. In addition to the hunting of stationary game, the hominids must have operated most effectively against prey of small to moderate size, since there is no evidence to favor the view that they were technologically aided by effective projectiles or heavy shock weapons that would make operations against large prey feasible. It is true that small predators are capable of taking prey substantially larger then themselves if they are social hunters, such as the Cape hunting dogs or spotted hyaena. However, the data presented here clearly indicates a bias in favor of the essentially nocturnal, solitary, and small antelope when they appear in the faunal assemblage in "meat biased" frequencies; that is adults that were butchered and transported in terms of a biased selection of meat-yielding parts. These animals have a modal body size of approximately 18 kg (40 pounds). This fact alone seems sufficient to eliminate the idea of hominid "pack hunting" and/or effective cooperative hunting with shock weapons.

At least during MSA times, the hominids seem to have been taking prey that were (1) essentially stationary, (2) small in size (about 18 kg), and (3) scattered and dispersed in brush and scrub cover during the day. These facts are most consistent with a strategy that requires the hominid to catch and hold the prey while it is killed. Killing and wounding prey with fang and claw while the prey is still mobile does not seem to be a realistic behavioral model, given the smaller size of the prey relative to the body size of the hominids, which was certainly above 18 kg.

The behavior of the leopard is interesting to consider relative to the record of predation seemingly indicated at Klasies River Mouth. The leopard is essentially a solitary killer, taking game primarily by stealth and skill in stalking and ambush. It may vary in weight between 50 and 80 kg (110 and 180 pounds). George Schaller describes nicely the prey of the leopard:

> Among the large prey items were Thomson's gazelle which weigh at most 23 kg, reedbuck 66 kg, impala 64 kg, and Grant's gazelle 70 kg. . . . In addition, a female topi and hartebeest weighing about 109 and 126 kg respectively, were killed, as was a yearling male wildebeest weighing an estimated 130 kg. Kruuk and Turner (1967) reported a yearling topi, a yearling female wildebeest and an adult female wildebeest among their large kills. Other prey consisted of small species or the young of large ones. Leopards seem to prefer prey in the 20 to 70 kg size category with an upper limit at about 150 kg, two to three times the weight of the cat itself. Seventy-eight percent of the prey in Kruger Park consisted of impala (64 kg) and Pienaar (1969) noted that most of it does not exceed the weight of a leopard. (Schaller 1972b:291; © 1972 by the University of Chicago)

We might generalize that the leopard is most effective killing prey around, or only slightly under, its own body weight.

Based on the data from Klasies River Mouth, the animal most commonly taken by the hominids for meat was the Cape grysbok, which weighs between 20 and 30 pounds, or about 16 kg. I think we can expect the hominids involved to have been at least as large as the leopard. Their capture–kill capacity seems, however, to have been considerably less effective than that of the leopard. Physically overpowering the prey prior to killing it seems to be the best way of visualizing the capturing tactics of the hominids operating out of Klasies River Mouth Cave 1.

The Klasies hominids seem primarily to have been killing stationary animals of a size sufficiently small for them physically to overpower and hold the prey prior to killing. What they hunted then, were *opportunities* to kill. They did not hunt particular species as such. Searching for opportunities ensured that they would regularly kill the young of large species if they fell in the size and situational context of a killing opportunity.

In this light, an interesting observation has been made regarding the basic contrasts in infant–mother relationships among the ungulates (Lent 1974). One basic pattern is called the *hide-a-baby* strategy, in which the mother gives birth in solitude and generally in fairly dense vegetation. After birth, the young are left hiding in the vegetation while the mother goes out feeding, returning at intervals to nurse the newborn calf. This behavior has been well described by Spinage:

> The waterbuck fawn is a "hider," remains hidden during the first 2 to 4 weeks; 3 weeks being the average period which I observed. This concealment is in a circumscribed area, but the fawn does not have a regular "form." It simply hides itself in the nearest long grass, or thicket when the dam leaves it. During the period of lying-out the fawn is visited only once in 12 hours, and probably only once during the whole 24 hours. (Spinage 1982:122)

In marked contrast to the hide-a-baby strategy is the *walk-along* strategy, in which very shortly after birth the young fawn begins to follow its mother as she moves about feeding. The hide-a-baby strategy is most common among the ungulates of smaller body size and/or the more territorial forms that inhabit forest and scrub-brush habitats. Forms inhabiting open habitats, such as the alcelaphines (wildebeest, hartebeest, topi, etc.), practice walk-along strategies.

As I have suggested, the hominids were hunting for opportunities, and these were relatively small stationary prey. The very young of the ungulates practicing a hide-a-baby mother–young relationship would certainly present themselves to the hominids as opportunities. In general, these appropriate-size animals would most likely occur in the same types of woodland–brush habitats in which the hominids took the small adult bovids such as the grysbok.

Seen in this light, the correlation between size of adult and frequency of unworn milk teeth summarized in Table 5.2 makes even more sense. The young of the larger species, such as the 4000-pound (1814 kg) adult *Pelorovis*, would only be in the appropriate "prey-size window" for the hominids when their young were first born, whereas the young of smaller species may have been within the size range of potential prey longer, during their early maturational period. I would agree, then, with Klein's argument that the hominids of Klasies River were basicially hunting the very young of a variety of species.

I think it is fair to generalize that modern men capture animals primarily with the aid of traps and game surrounds. With both of these strategies modern man is not dependent primarily upon his own physical strength to capture the prey. Therefore I suggest that the evidence for the taking of live prey at Klasies is biased in favor of small-body-size animals relative to the body size of the hominid. Viewed against the leopard's behavior, this bias is consistent with a model of the hominid first capturing prey, then overpowering it with his unaided physical strength, and afterward killing the captured animal. In addition, the hominid may have been actually running away from irate mothers after having captured their young. Under such conditions, the speed of the hominid would be important only *after* capturing prey. This would certainly favor the capture of small prey and taking it in cover, which would conceal the hominids from aroused mothers. It might even be possible that early hominids escaped with small prey into trees. Certainly such behavior would be hindered as a direct function of increasing prey sizes, and as we have been, the hominids seem to have been systematically taking small prey relative to their body size.

The capture-carry-and-kill model of predation suggested here accommodates the small body sizes characteristic of both the young of large species, and the adults of small species. On the other hand, direct capture without carrying off the prey would seemingly be more likely in situations in which the adults of small species were captured. As a further clue to hominid hunting tactics, we need to study the size differences between the young of strict hide-a-baby species versus the adults of the small nocturnal species.

Hunting and Age Profiles

As in the cases already discussed, Klein's interpretations of Klasies data have assumed hunting to be the tactical means whereby all the faunal remains ended up in the Klasies River Mouth site. This is true for the very large animals as well as the small animals. He has interpreted the differences noted in his studies between species, particularly in age profiles, as directly reflecting differing hunting tactics. He describes a "catastrophic" age profile for eland (*Taurotragus*) and bushbuck (*Tragelaphus scriptus*) (Klein 1981: Figure 5). He suggests that this is evidence of animal drives and the use of snares by the hominid occupants of Klasies River Mouth.

I have already suggested that the eland were in the Klasies River Mouth

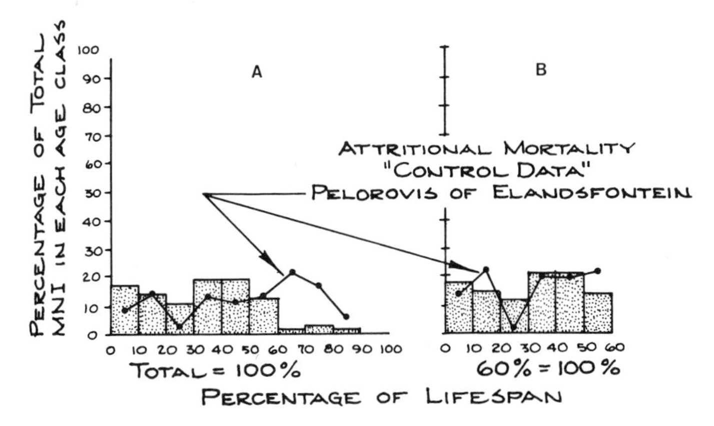

Figure 5.7 Comparison between eland (*Taurotragus*) age (mortality) profiles from (A) Klasies River Mouth and (B) the attritional profile for *Pelorovis* at Elandsfontein used as a control by Klein (1978c).

site because the hominids scavenged the carcasses of nocturnal kills around water sources, perhaps by lions. Figure 5.7 compares the age profile of the eland from Klasies Cave 1 with the age profile presented by Klein as representative of an attritional pattern, for *Pelorovis* from the site of Elandsfontein. What we note is that there are more young eland at Klasies than one would expect in a normal attritional-death profile. We saw the same pattern of underrepresented old, and inflated prime and young *Pelorovis* remains at Klasies River (see Figure 5.5), in spite of the fact that Klein has suggested that it was an attritional pattern resulting from human hunting of animals in all age classes, but experiencing differential success among the young and old categories. The data, however, do not support this view. In both *Pelorovis* and *Taurotragus* (eland) there is a marked underrepresentation of old individuals. I have already suggested that this represents a bias against large head-parts and favoring younger and smaller head-parts for transport back to the site at Klasies River Mouth. Put another way, heads are primarily exploited for their meat and fat. Fresh carcasses, where such edible materials were available, were primarily young individuals and in many cases they were very young indeed, being newborn animals that had been killed by the hominids.

The size bias in hunting thus ensures an age bias favoring young of large species exploited for meat. On the other hand, my data on the relative frequencies of fused versus unfused lower-limb bones from the species being discussed (Figure 5.3) show a bias in the other direction, favoring fully adult or even old individuals. A discrepancy between the age profiles for teeth versus age estimates based on epiphyseal fusion is believed to arise from tactics standing behind the exploitation of heads versus the lower limbs. At Klasies, heads are biased in favor of young animals, lower limbs are biased in favor of older animals. That derives from the hunting of relatively helpless young and the scavenging of marginal parts (marrow bones) from the ravaged death sites of larger animals already fed upon by other predators and/or scavengers.

Let us return to the argument presented by Klein, namely, that eland were represented by more adults and hence constituted a catastrophic age profile indicative of mass killing, perhaps even animal drives by the hominids (Klein 1978c:213; 1982:153). We have already seen that there is a marked bias against old individuals for the parts of the head. It remains to determine if the pattern in the young and young-adult eland do in fact differ from a normal attritional-age profile. This question may be answered by comparing the younger half of the eland age graph—that is, the section that represents animals that have completed up to 60% of their life span—with a comparable segment of the control example of an attritional-age profile as presented by Klein. I calculated a chi-square value for the comparison between the control age profile of *Pelorovis* from Elandsfortein (Klein 1982) and the eland data from Klasies River Mouth (Klein 1978c:205) and found a summary chi-square value of 5.92 for five degrees of freedom. A value this large could be expected to arise by chance alone at least 40 times out of 100 random events. By the conventions normally used by statisticians, there is no significant difference between the eland frequencies and the control "attritional" data as presented by Klein for *Pelorovis* from Elandsfontein. In other words, the eland graph from Klasies River Mouth is a normal attritional graph biased against old animals, as was also the case for the buffalo remains from the same site. The only difference between the graphs is the inflated frequency of very young buffalo represented at Klasies River Mouth. This has already been suggested to arise because of the killing of the very young buffalo by Klasies hominids. It is interesting that the same age bias favoring very young animals, does not typify the eland or the wildebeest, although it is clearly characteristic of the giant buffalo, Cape buffalo, and bastard hartebeest. The single factor that seems to distinguish the two sets of animals is that both the eland and wildebeest are best described as extremely nomadic, moving over vast areas of changing pasture. On the other hand, all the other animals listed are relatively territorial and are even sometimes described as being rather sedentary.

The eland, which are notoriously mobile, have been described as generally giving birth just shortly after the peak rains during the flush of new grass (Moss 1975:183). In correlated fashion they have been described as occurring in large herds during the wet season when they feed off the new grass. Then they break up into smaller feeding units dispersing into thicker vegetation during the dry months, when they browse off the brush and trees of the scrub-vegetation zones (Kingdon 1982:129). Of course, during the latter months they would not be giving birth. Certainly the sparse grass and the scrub vegetation of the *fynbos* plant community characteristic of the area around Klasies River Cave 1 is likely to have been attractive to eland only during the dry season, when they were dispersed. This would also have been the time when births would be least likely; hence no hunting of young by hominids, but the successful ambushing of eland by lion around relatively localized water sources. The same grass-seeking behavior is characteristic of wildebeest (see Sinclair 1977:177), and as in the case of eland the young tend to be born during the peak of the grass production, following the rains. Given that wildebeest would not be likely to enter the *fynbos* scrub vegetation seeking grass, it is also unlikely that they would enter the zone at the time of calving. In both of these cases, the bias at the site of Klasies (given hominids who hunted young animals) is against the very young, who are unlikely to be seasonally present in the Klasies setting. Calves of the year are, however, represented among the smaller transient species.

On the other hand, all the species exhibiting inflated frequencies of very young individuals are species that are generally nonnomadic and tend toward territorial behavior. This means that they would be year-round residents of the Klasies region and could be expected to calve in the area. When they did, the hominids apparently took advantage of their relatively helpless young.

The difference between the age profiles of eland and wildebeest versus giant and Cape buffalo, as well as bastard hartebeest, is not that the former were hunted with mass drives and corporate hunting tactics (yielding a catastrophic age profile), whereas the latter were hunted individually yielding an attritional-age profile (Klein 1979:158–159) with a bias in favor of very young. This analysis suggests instead that (1) there is a strong bias in all the age profiles of large animals against very old individuals, distinguishing them from the attritional profile as documented by Klein for giant buffalo at Elandsfortein; (2) there is a very real bias favoring very young individuals among the Cape buffalo, giant buffalo, and bastard hartebeest; and (3) there is no statistical difference between the age profile for eland and the control attritional age profile for giant buffalo in the age categories from 0 to 60% of the life span. Stated another way, eliminating the bias against old individuals, there is no difference between the eland graph and a common attritional profile!

What must be explained, therefore, is not why one set of animals that exhibit a catastrophic age profile (eland and wildebeest) and another that exhibit an attritional profile (Cape and giant buffalo plus bastard hartebeest) are biased in favor of very young animals, as Klein has assumed. The challenge is to explain why all the large animals are biased against very old individuals, and why in addition the buffalo–bastard-hartebeest group is further biased in favor of many very young individuals, while all age profiles represent varying forms of an attritional pattern.

The point emphasized here is that a single species may be exploited with different tactics. In addition, the same tactics—when used on animals of different species, sex, size, and age—may result in very different faunal assemblages at a given place (see Binford 1982b).

Thus far this discussion of hunting has focused on the differential success of the hominids in killing animals of different size regardless of species, as well as the different access that hominids had to the young of transient species if they were resident in the *fynbos* biome. Both of these forms of emphasis have taken individual hunting for granted—that is, the taking of animals one at a time. As previously noted, Klein (1978c:213; 1982:153) has suggested that different hunting tactics were perhaps involved. Is there some way of positively evaluating the alternative argument that the differences were not related to different access or success of a tactic (given different species' characteristics) but to different hunting tactics? Certainly the study of kill sites, hunting facilities, and the overall pattern of the archaeological remains of subsistence is a very important direction for future research. But now, in the absence of such direct evidence of hunting tactics, is there some way we can use the available data to assess the probability that the hominids of Klasies River Mouth practiced hunting tactics that resulted in the killing of multiple animals—or making mass kills?

Long ago I measured the crown heights on horse teeth from the Mousterian site of Combe Grenal. I never standardize the measurements relative to controlled life spans, but I think it is clear that the shorter, fully erupted adult teeth are the more worn and hence represent older animals at death. Table 5.3 summarizes the crown-height data for horse upper second and third molars versus the lower, or mandibular, second and third molars recovered from the denticulate Mousterian (G and H) Levels 14–16 at Combe Grenal.

What is well illustrated is that 17 (71%) of the lower-rear teeth are from old individuals, whereas only 15 (39%) of the upper teeth are from old individuals. This betrays an age bias in returning parts of the head to the site. When crania with maxillary teeth were returned, they were almost always from young individuals.

I observed this same bias among the Nunamiut Eskimo, among whom the heads of adult animals were judged to be tough and to yield little food relative to their size. On the other hand, mandibles with their attached tongues taken from adult animals were considered a treat and were regular-

TABLE 5.3

Crown Heights for Horse Teeth from the Mousterian Site of Combe Grenal[a]

	Old				*Young*				
	2–2.9	*3–3.9*	*4–4.9*	*Subtotal*	*5–5.9*	*6–6.9*	*7–7.9*	*Subtotal*	*Total*
Lower	2	12	3	17	1	3	3	7	24
Upper	3	7	5	15	6	12	5	23	38

[a] In mm.

ly returned because of their larger size. This same pattern seems to have characterized the behavior of the Neanderthals at the site of Combe Grenal. Such an age bias among head parts, however, is definitely not present in the Klasies remains. Richard Klein (1978:203) has presented interesting data on the relative frequencies of milk dentition (indicative of young animals) as well as third molars, which are the last molars to erupt and hence are indicative of prime-age to old individuals, for both mandibular and maxillary teeth. These data make it possible to evaluate whether there is a differential age bias for maxillary–cranial parts versus mandibular–tongue parts. Table 5.4 summarizes these data from Klein.

It is very clear from Table 5.4 that there is no age bias between upper and lower teeth, as was seen in the horse remains from Combe Grenal. In all cases at Kalsies there is a strong bias in favor of lower teeth (approximately two mandibles were introduced for every cranium). There was no age bias between the introduced heads with attached mandibles versus the isolated mandibles. Isolated mandibles were drawn from the same age population as

TABLE 5.4

Crown Heights of Several Species from Klasies River Mouth[a,b]

	Pelorovis			*Cape buffalo*			*Eland*		
	Young	*Old*	*Total*	*Young*	*Old*	*Total*	*Young*	*Old*	*Total*
Lower	44	26	70	32	9	41	24	99	123
	(.63)	(.37)	(.68)	(.78)	(.22)	(.65)	(.20)	(.80)	(.62)
Upper	20	13	33	17	5	22	16	60	76
	(.61)	(.39)		(.77)	(.23)		(.21)	(.79)	
	64	39	103	49	14	63	40	159	199

[a] *Young* is represented by milk dentition of the fourth premolar, and *old* is represented by adult third molars. Data from Klein (1978: Table 4).

[b] In mm. Values in parentheses are percentages of young and old within each anatomical class.

were the heads. This suggests to me that all head parts, both crania and mandibles, were introduced for their meat and that whether or not heads were introduced may have been a matter of (1) distance from site, and (2) size of animal. There is a regular reduction of bias in favor of mandibles as we go down the body-size sequence from giant buffalo (68%) to eland (62%). The lack of age bias between head parts may simply reflect the lack of options. If I am faced with an old animal and a young animal and can only carry one animal, I may then choose the young, more tender animal head.

On the other hand, if I am rarely faced with a situation in which I have multiple animals available at one time, then I can only make choices among parts anatomically differentiated within a single animal. I can choose the tongue as a more desirable part vis-à-vis the head, but I cannot choose a young instead of an old tongue. It is true that if I have a long-term perspective on the procurement of foods I could differentially pass up opportunities as they present themselves, choosing to exploit a young animal when encountered—taking, for instance, its head—but at another time passing up the head of an old animal, taking only its mandible. I could get an age bias among anatomical parts if (1) I had multiple animals of different ages available for exploitation at the same time (forming the population in terms of which choices were made), or (2) there was considerable time depth to the decision-making framework providing a temporally "deep" set of opportunities among which choices could be made. The absence of age bias between mandibles and crania for all the species for which data are available is taken as strong evidence that (1) animals were exploited by the Klasies hominids one at a time (there were no mass kills or death populations available for exploitation), and (2) that the planning depth in terms of which the hominids exploited carcasses was very shallow, being essentially a "thing of the moment." Decisions were made in terms of immediate contingencies, such as distance from site, state of the carcass, and numbers of persons exploiting the carcass, and not relative to potential opportunities not immediately apparent.

This conclusion buffers the argument previously made regarding the lack of a catastrophic age profile for eland and wildebeest. I think that I have accounted for the patterns of differential age represented among the species discussed by Klein. The assumptions made by Klein—namely, that there was no "human bias" (Klein 1978c:203–204) and that all the remains were hunted by hominids (Klein 1976:87; 1980:229–230)—does not appear justified. Similarly, the postulation of mass animal drives to account for the eland data (Klein 1978; 1982:153) does not appear warranted, nor does the suggestion that the "catastrophic mortality profile" of the small antelope imply the use of snares. The little animals are small enough to be taken as adults by a predator the size of the hominids even if the capture is technologically unaided. Finally, the case for the hunting of the young of the large buffalo—both the Cape buffalo and the giant buffalo (*Pelorvis*)—seems supported, whereas scavenging the adult forms is almost certainly a characteristic of the Klasies hominids.

Instead of the interpretations offered by Klein, I think it is much more likely that the hominids were prosecuting a single set of food-procurement strategies, undifferentiated tactically among the species. They were (1) hunting very-small-body-size individuals (including the young of large species when they were available), and (2) scavenging marginal food remains from the ravaged death sites of animals with a bias for lion kills around water sources during the dry season.

The differences among the age profiles of the animals introduced to Klasies River Mouth Cave 1 did not arise from differing species-specific hunting tactics as suggested by Klein, but instead from different success rates for the same tactics, tempered by different behavior patterns of different sets of species. It appears that there was a "transient" set represented by eland, wildebeest, and probably the other alcelaphines as well, which appeared in the coastal *fynbos* during the dry season, perhaps coming from the interior valleys. Grass feeders commonly disperse into higher-rainfall scrub areas when the grasslands are dry and overgrazed. It is quite possible that the exploitation of these species is a good seasonal indicator, as well as perhaps providing clues to environmental changes during the course of the archaeological accumulations. The other group of species represents the resident *fynbos* set of animals. These include all the small animals such as hyrax, small antelope such as the Cape grysbok, medium-size browsing antelope such as the bushbuck and kudu, together with the giant buffalo and Cape buffalo. The larger of the resident species were exploited as young during their calving period, with the smaller species exploited more continuously and commonly as adults because the hominids were apparently able to take handily animals up to about 90 pounds (41 kg). The medium-size resident forms appear to have been primarily scavenged, whereas some the medium-size transient species were exploited while young.

Implications of Variable Subsistence Tactics for Environmental Reconstruction and Dating

In Chapter 2 I described the environment of Klasies River Mouth and summarized some of the arguments that have been advanced regarding the age of the site and its importance for understanding human evolution. One issue that tempered the ways in which the faunal facts had been used for inferring environments of the Upper Pleistocene was the assumption noted above that hominids hunted the animals recovered from the archaeological sites. It was also generally assumed that patterned changes in species variability among stratigraphic units within sites was a direct reflection of climatically induced changes in the animals available for the hominids to use (see Klein 1980:253).

I think I have been successful in mounting a robust argument to the effect that hominids both hunted and scavenged. Therefore the bones present in many archaeological sites are not a simple function of the pursuit of a single food-procurement strategy—for example, hunting. Given that at least two major strategies were involved and that hunting was a dominant strategy relative to scavenging among recent hunter–gatherer occupants of the area (see Figure 5.8), we may reasonably ask, (1) When and under what conditions did hunting become dominant? and (2) What differential effect do shifts in strategies—hunting versus scavenging—have for relative species frequencies in sites that span such strategy shifts? If such shifts could be demonstrated, then the current tendency to interpret frequency shifts among species at archaeological sites as a direct measure of climatic shifts would be certainly suspect.

> The percentages of grassland creatures (equids and alcelaphines) and of creatures preferring more closed habitats (especially antelopes of the genera *Tragelaphus, Raphicerus,* and *Cephalophus*) are roughly comparable between the *LSA* fauna and that from the oldest MSA culture stage, MSA I . . . . In the MSA stage II there is an apparent decrease in open country forms which may indicate an increase in the amount of closed vegetation. . . . there is a marked (and statistically significant) increase in *Raphicerus* and *Tragelaphus* and a corresponding decrease in open country forms in the levels immediately overlying 38/39 in Cave 1. (Klein 1976:78)

What is interesting in Klein's statement is that he makes it very explicit that the majority of the species taken as indicators of grasslands are relatively large: wildebeest, hartebeest, bastard hartebeest. These are all medium–large-body-size animals of size class III as used in this study. On the other hand, two of the three groups mentioned as indicative of bush cover—*Cephalophus* (blue duiker), and *Raphicerus* (Cape grysbok)—are both in the smallest size class (I) of animals, while the third, *Tragalaphus,* is represented by bushbuck, a small–medium-size class II animal, and by kudu (a generally rare species), which is equal in size (class III) to the species considered indicative of environments where grass is more plentiful. In effect, the species taken as indicative of bush cover are small, and as we have seen were generally hunted or at least introduced to Klasies River Mouth Cave 1 in terms of meat considerations. On the other hand, the species listed as indica-

Figure 5.8 Tribesmen driving lions from a carcass for the purpose of scavenging. The watercolor drawing was made in 1835 by C. D. Bell.

tive of grassland settings are all generally larger and are those that we have seen to have been primarily scavenged, except for their young. Clearly, shifts in what appear to be climate as monitored by the "bushy" species could equally reflect a *behavioral* shift in favor of more hunted foods with no necessary changes in environment. This means that it is reasonable to seek ways of monitoring each possibility, behavior and environment, independently of the other.

I have reproduced data from Tables 1, 2, and 3 of Klein's (1976) study of the Klasies River Mouth fauna, so that the basic data discussed here are readily available, and the same data used by those making climatic inferences (Tables 5.5–5.7).

Ideally I would like to hold body size constant and then compare species that were apt to have been hunted (small) with those that were apt to have been scavenged (large). Because there is a strong autocorrelation between body size and procurement strategy, my approach must seek to hold habitat preference constant. For instance, I might look at animals who are considered grassland species and then compare species within each set that differ markedly in body size across stratigraphic time. These chrono-stratigraphic comparisons would seek to determine whether there is any justification for the idea that food-procurement strategies might proportionally vary through time. After we know the nature of such patterning, we could

TABLE 5.5

The Minimum Numbers of Individuals by Which Each Mammalian Species Is Represented in the Various Horizons of Klasies River Mouth Cave[a]

	LSA II	LSA I	MSA IV	MSA II					MSA I	
	1–6	7–12	13	14	15	16	17a	17b	37	38/39
Homo sapiens, Man				½	1	1	½		?1	?1
Papio ursinus, Chacma baboon	2		1	3	1	1			1	
Canis mesomelas, Black-backed jackal		1					1			
Mellivora capensis, Honey badger				1		1				
Aonyx capensis, Clawless Otter	2			2	1	2	1	1	1	
Genetta sp., Genet	?1									
Herpestes ichneumon, Egyptian mongoose				1	1	1				
H. pulverulentus, Cape grey mongoose				2		1	1	1		
Atilax paludinosus, Water mongoose				1						1
Hyaena brunnea, Brown hyena				1	1	1				
Felis libyca, Wildcat						1				
Felis cf. *caracal*, Caracal				1	1		1			
Panthera pardus, Leopard	1			4	1	1	1	2	1	
Arctocephalus pusillus, Cape fur seal	7	8	3	20	5	17	4	4	7	4
Mirounga leonina, Elephant seal				1						
Loxodonta africana, Elephant						1	1	1		
Procavia capensis, Rock hyrax		2	3	15	5	15	3	6	2	2
Diceros bicornis, Black rhinoceros				2		1	1		1	

Equus cf. *quagga,* Quagga				1			1		1	1
Potamochoerus porcus, Bushpig	?1			1	1				2	2
Phacochoerus aethiopicus, Warthog					1					2
Hippopotamus amphibius, Hippopotamus	1	2		4	1	2	1	2	5	1
Cephalophus monticola, Blue duiker	2	3								
Raphicerus melanotis, Cape grysbok	1	7		21	14	5	6	3	4	
Ourebia ourebi, Oribi						1	1			
Pelea capreolus, Vaalribbok	2	2	2	1		3	1			
Redunca cf. *arundinum,* Southern reedbuck			1	1	1		1		2	2
R. falvorufula, Mountain reedbuck	1	1				1	2		3	1
Hippotragus leucophaeus, Blue antelope			4	8	4	6	7	7	11	5
Alcelaphus buselaphus, Hartebeest	5	3	2						1	2
Damaliscus sp., Bastard hartebeest				1		1				
Connochaetes sp., Wildebeest				1	1	2	2	2		5
Antidorcas sp., Springbok				1		2	1			
Tragelaphus scriptus, Bushbuck			1	6	8	2	3	1	1	2
T. strepsiceros, Kudu				2		5	1	2	3	
Taurotragus oryx, Eland		1	3	27	10	23	12	8	10	11
Syncerus caffer, Cape buffalo	7	7	4	5	3	9	4	8	7	4
Pelorovis antiquus, Giant buffalo			2	13	1	9	4	5	11	7
Hystrix africae-australis, Porcupine	3	1	1	10	4	3	1		2*	1
Georychus capensis, Mole rat				2	3	1	1			
Lepus capensis, Cape hare				1						
Delphinidae, Dolphins		2	1	2	2	1	1	1	2	1
Other Cetacea, Whales				1					1	?1

[a] From Klein 1976:Table 1.

TABLE 5.6

The Minimum Numbers of Individuals by Which Each Mammalian Species Is Represented in the Various Horizons of Klasies River Mouth Cave 1A[a]

	MSA III					*Howieson's Poort*			*MSA II*									
	1–3	*4*	*5*	*6*	*7–9*	*10–11*	*13–16*	*17–21*	*22*	*23–24*	*25*	*26*	*27*	*28–29*	*30*	*31*	*32–33*	*34*
Homo sapiens, Man				1		?1					?1							
Papio ursinus, Chacma baboon					2		2	1										
Herpestes pulverulentus, Cape grey mongoose								2										
Atilax paludinosus, Water mongoose				1														
Panthera pardus, Leopard				1														
Felis libyca, Wildcat						1		1	1									
Felis cf. *caracal,* Caracal					1		1											
Arctocephalus pusillus, Cape fur seal			1	3	3	2	5	6	1	3	2		2	4	2	2	5	2
Loxodonta africana, Elephant								1										

Procavia capensis, Rock hyrax	4	1	2	3	3	4	4	4		1	1	1	1	1			2	1
Equus cf. *quagga,* Quagga					2	3	3	6										
Hippopotamus amphibius, Hippopotamus				1			1	2										
Raphicerus melanotis, Cape grysbok	2			1	1	1	6	2					1	1	1	1	1	
Pelea capreolus, Vaalribbok				1			2										1	
Redunca cf. *arundinum,* Southern reedbuck					1	1	1	3		1	1		1					
Hippotragus leucophaeus, Blue antelope	2			1		1	5	3	1	2		1	1	2		1		
Damaliscus sp., Bastard hartebeest				1		3												
Connochaetes sp., Wildebeest					1	3		1										
Tragelaphus scriptus, Bushbuck								1										
T. strepsiceros, Kudu								1			2			1			3	
Taurotragus oryx, Eland	1		3	1	3	3	1	7		2		1	2	6	1			
Syncerus caffer, Cape buffalo	2		1	4	1	6	2	4		3		1	1					1
Pelorovis antiquus, Giant buffalo				1	1	5		2	1	1			2					
Hystrix africae-australis, Porcupine							1	1					1				1	
Delphinidae, Dolphins						?1												

[a] From Klein 1976:Table 2.

TABLE 5.7

The Minimum Numbers of Individuals by Which Each Mammalian Species Is Represented in the Various Horizons of Klasies River Mouth Caves 1B[a]

	KRM 1B–MSA 1												
	1–3	*4*	*5*	*6*	*7*	*8*	*9*	*10*	*11*	*12*	*13*	*14*	*15*
Homo sapiens, Man								1					
Papio ursinus, Chacma baboon	1												
Canis mesomelas, Black-backed jackal	1												
Atilax paludinosus, Water mongoose								1					
Arctocephalus pusillus, Cape fur seal	3	2	1	2	3	2	1	3	2		1		1
Procavia capensis, Rock hyrax	3		3	3	1			1	2		1		
Potamochoerus porcus, Bushpig											1		
Hippopotamus amphibius, Hippopotamus				?1							1		
Raphicerus melanotis, Cape grysbok	4		1		2								
Pelea capreolus, Vaalribbok			1										
Hippotragus leucophaeus, Blue antelope				2				1					
Alcephalus buselaphus, Hartebeest						1	1						
Connochaetes sp., Wildebeest						1		1					
Tragelaphus scriptus, Bushbuck				1									
Taurotragus oryx, Eland	4	1		1		2	1						
Syncerus caffer, Cape buffalo	1												
Pelorovis antiquus, Giant buffalo				1									
Hystrix africae-australis, Porcupine				1									
Lepus capensis, Hare										1			
Delphinidae, Dolphins													

[a] From Klein 1976:Table 3.

then seek to determine if there was coincident variability that was also referable to changing environments.

For the first comparison I have chosen three species. The Cape grysbok is in the smallest body-size class (class I). Earlier analysis suggests this animal was being hunted almost exclusively. This species is considered endemic to the Cape biotic province (Jarvis 1979) and should therefore be a good indicator of *fynbos*-type cover relative to grasslands. Similarly the blue antelope is generally considered to be endemic to the Cape biotic province (Klein 1980:251), and is therefore adapted to the closed-to-broken cover of the *fynbos*-type biome. The estimated body weight of the blue antelope is between 200 and 300 pounds (91–136 kg), placing it in body-size class III. The analysis of the frequency of fused bones (see Figure 5.3) suggested that these animals exhibited an age profile similar to eland—therefore strongly suggestive of having been scavenged. The actual frequencies of anatomical parts also is completely consistent with the arguments for recognizing scavenging presented here. Finally, the eland (*Taurotragus*) is a mixed feeder and might be found in grasslands, while during dry season it may occur in areas supportive of more "bushy" vegetation and more permanent water. This species was chosen because it is clearly representative of a biased exploitation by scavenging, and there is little to support the view that its young were exploited as live animals (see the discussions of age profiles and refer to Figure 5.7).

The pertinent quantitative information on these species is summarized as Inventories I and II of Table 5.8. These data are plotted in Figure 5.9, arranged chronostratigraphically from bottom to top, earliest to latest. It is obvious that the percentage of Cape grysbok relative to the Cape-endemic blue antelope steadily increases from the basal Level 38 up through Level 15. Put another way, a regular temporal trend is clearly indicated, showing a steady increase in the frequency with which the small, *hunted* Cape grysbok was being taken relative to the blue antelope (thought to have been scavenged) until Levels 13 and 14 are encountered. Level 13 is the only level interpreted by Butzer as having been accumulated during a major low-water stage, when grassland species would be expected to be more common. Level 14 is the one with all the "other" agents represented and was also likely to have been eroded. It will be recalled all the other levels were interpreted as having accumulated during high sea-level stages and hence represent episodes when *fynbos* types of (bushy) vegetation would have been most common on the Klasies coast.

The sequence of high-water stages that have been equated with the various levels at Klasies Cave 1 are all substages in the overall isotopic Stage 5 thought to represent the early Upper Pleistocene (125,000–60,000 B.P.). Ironically, if one projects the ocean levels to terrestrial temperatures, the

TABLE 5.8

Inventories Abstracted from Klein's Basic Data Tables[a]

	Cave 1									Shelter IA			Shelter IB
Inventory	*38*	*37*	*17B*	*17A*	*16*	*15*	*14*	*13*	*LSA*	*MSA II*	*HP*	*MSA III*	*MSA I*
I													
1. Eland	11	10	8	12	23	10	27	3	1	8	11	18	9
2. Grysbok	0	4	3	6	5	14	21	0	8	4	9	6	7
3. Total	11	14	11	18	28	24	48	3	9	12	20	24	16
4. % Eland	100	71	73	67	82	42	56	100	89	67	55	75	56
II													
5. Blue antelope	5	11	7	7	6	4	8	4	2	11	9	3	3
6. Grysbok	0	4	3	6	5	14	21	0	8	6	9	4	7
7. Total	5	15	10	13	11	18	29	4	10	17	18	7	10
8. % Grysbok	0	27	33	46	45	78	27	0	80	35	50	57	70
IIIa													
9. Grysbok	0	4	3	6	5	14	21	0	8	4	9	6	7
10. Bushbuck	2	1	1	3	2	8	6	1	2	0	1	0	1
11. Subtotal	2	5	4	9	7	22	27	1	10	4	10	6	8

IIIb													
12. Kudu	0	3	2	1	5	0	2	0	0	3	1	0	0
13. Blue antelope	5	11	7	7	6	4	8	4	2	11	9	3	3
14. Subtotal	5	14	9	8	11	4	10	4	2	14	10	3	3
15. Grand total	7	19	13	17	18	26	37	5	12	18	20	9	11
16. % Small (IIIa)	29	26	31	53	64	85	73	20	83	22	50	67	73
IV													
17. Hartebeest	2	1						2	8				2
18. Wildebeest	5		2	2	2	1	1				4	1	2
19. Bastard hartebeest						1	1				3	1	
20. Subtotal	7	1	2	2	2	2	2	2	8	0	7	2	4
21. Total IIIb and IV	12	15	11	10	13	6	12	6	10	14	17	5	17
22. % IIIb (Bushy)	42	93	82	80	85	67	83	67	20	100	59	60	43
V													
23. Eland	11	10	8	12	23	10	27	3	1	8	11	18	9
24. Part IIIa total	2	5	4	9	7	22	27	1	10	4	10	6	8
25. Total	13	15	12	21	30	32	54	4	11	12	21	24	17
26. % Eland	85	67	67	57	77	31	50	75	09	67	52	75	53

[a] See Tables 5.6–5.8.

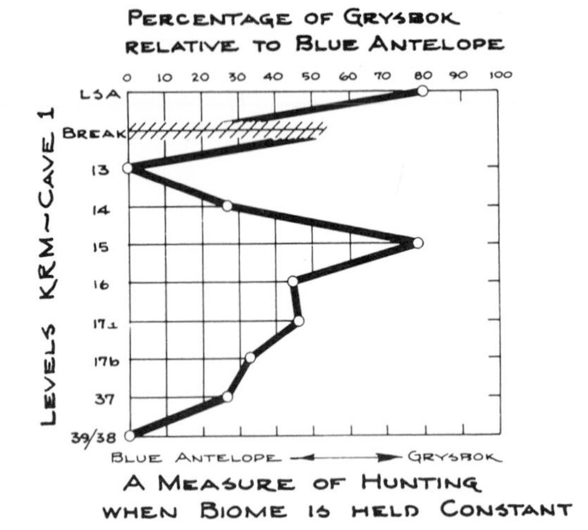

Figure 5.9 Comparison between the frequencies of grysbok and blue antelope among the various levels from Cave 1, Klasies River Mouth.

overall warmth of the environment should be decreasing during this sequence and, hence, the interglacial optima for the Mediterranian type of *fynbos* vegetation should be correspondingly deteriorating (see Figure 5.09). Nevertheless, we see an increase in Cape grysbok—a *fynbos* endemic! This is only surprising if the levels are dated correctly, and if the "reverse" model of climatic change that was suggested by Klein (1972b), based on the fauna from the Nelson Bay Cave, is a correct picture of climatic dynamics associated with glacial (hypothermic) versus interglacial (hyperthermic) conditions in the Cape. That is, during glacial conditions grassland expanded southward, whereas during interglacials *fynbos* expanded northward into what is today *karoo*-type vegetation. If the grysbok is taken as an index to more closed cover as in *fynbos* or temperate forest conditions, then clearly (given correct dating) the *fynbos* appears to be expanding as the warm (hyperthermal) conditions are deteriorating. This is the reverse of the pattern inferred from the Nelson Bay data, yet is a pattern generally supporting the original zonal model of van Zinderen Bakker (1967:144). Clearly, things are complicated.

I think the graph if Figure 5.9 is best understood as in increase in the role of hunted foods and a corresponding decrease in the role of scavenged foods during the span of time represented by the Cave 1 deposits. As a check on the possibility that the blue antelope may not be a good monitor of scavenging strategies, Figure 5.10 shows the relationship between the frequency of Cape grysbok, considered as a hunted animal, and the eland, believed to be an unambiguous index to the role of scavenging. It is clear that the line on Figure 5.10 shows an almost perfect positive relationship

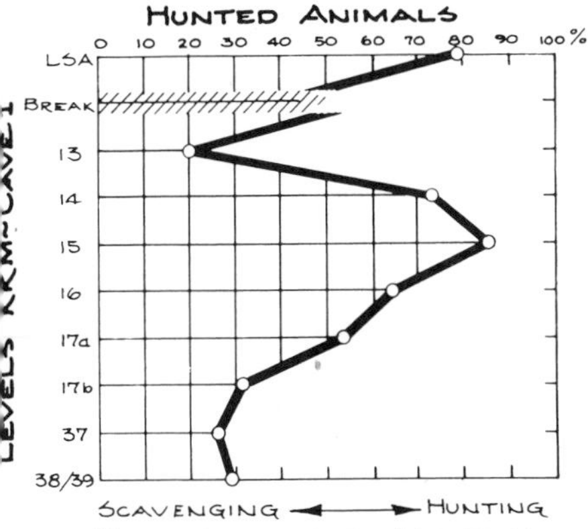

Figure 5.10 Comparison between grysbok and elands among the various levels from Cave 1, Klasies River Mouth.

with the line for grysbok versus blue antelope in Figure 5.9. Stated another way, the two large animals, blue antelope and eland, are positively correlated and hence both relate to the smaller grysbok in a similar fashion. The larger animals decrease at the expense of the smaller grysbok through the Klasies Cave 1 sequence until Level 13, in which the large forms again become dominant. Given the arguments developed earlier in this work, this means there was a regular increase in the relative role of hunting versus scavenging through the major part of the Klasies Cave 1 sequence with the trend only being reversed around Level 13. This pattern is perhaps even more striking when the frequencies of the small animals, such as grysbok and bushbuck, are combined and expressed against the frequencies of medium-body-size forms, the kudu and blue antelope, in Figure 5.11. All these are cover-loving forms and should be expected to vary in correlated fashion in response to environmental changes. Stated another way, in this comparison we are holding environment constant and looking at how body-size preferences vary through the Klasies deposits. We obtain a pattern very similar to that obtained when we compared only grysbok and blue antelope—namely,that what is varying are the preferences for animals of different body size, not animals at home in different environments.

This finding is even more forcefully demonstrated in Figure 5.11 where there are two different types of proportional frequencies shown. Relationship B traces the relationships between animals of medium body-size that are cover-loving species, versus those that are grassland-loving species. By holding body size roughly constant in this comparison, we obtain a very different picture of environmental change than was obtained by Klein

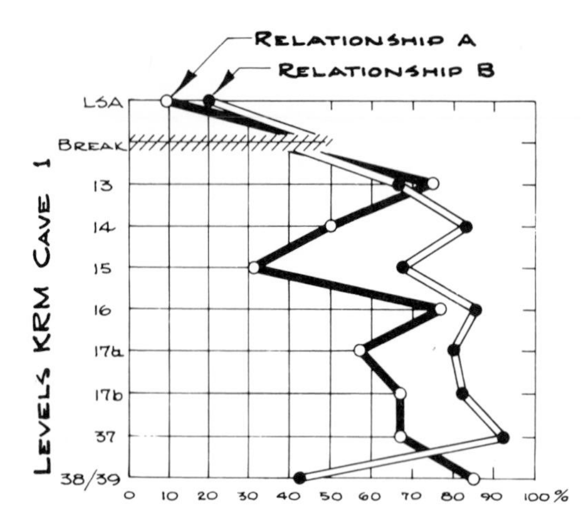

Figure 5.11 • Comparison between the frequencies of small antelope versus eland (Relationship A) and the percentage of medium-size antelope represented by bush-loving species (Relationship B) among the various levels from Cave 1, Klasies River Mouth.

(1976), because he did not control for body-size biases and saw them as reflecting environmental change (see Figure 2.12). We see a pattern of high levels of grass-loving species in Level 38 and then a relatively stable set of conditions indicated for Levels 37, 17b, 17a, 16, and 14, in which cover-loving animals dominate. In Levels 15 and 13 there were increases in grass-loving species and a marked increase shown in the LSA.

Several things make this pattern most provocative. If the levels at Klasies Cave 1 do in fact span a period of generally decreasing temperatures, then the direction of the pattern between Levels 37 and 13 is now in line with the model that expects increasing grassland-loving forms associated with colder environmental conditions. Such a pattern is consistent with the data from Nelson Bay Cave (Klein 1972b) and Elands Bay Cave (Parkinton 1972), in which both the dating and the environmental correlation seem secure. Given the assumed contemporaneity with a deteriorating environment or one moving in the direction of more glacially dominated climates (even during high-water stages), this pattern is now congruent with our other knowledge. This condition is essentially consistent with conditions during isotopic Stages 3, 4, or 5; or anytime during the Late Pleistocene prior to the glacial maximum, which is generally considered to have spanned the period between 27,000 and 18,000 B.P. Since most would agree that the levels under consideration date beyond the practical limits of ^{14}C-dating methods, this could be anytime between 32,000 and 128,000 B.P. (see Shackleton 1975:Table 1).

If we consider the additional points on the graph (Figure 5.11)—those for Level 38 indicate a marked dominence of grass-loving forms—we might expect these conditions during any low-water stage of the Late Pleistocene. The interesting and seemingly baffling point is the behavior of the LSA. As I have repeatedly noted, there has been a demonstration at Nelson Bay Cave, Elands Bay Cave, and others, that during the LSA there was an increase in cover-loving forms. This has been taken as coincident with the warming conditions at the close of the Pleistocene and the onset of the contemporary climatic regime. Inspection of Relationships A and B in Figure 5.11 shows that if one accepts the small antelope as indicative of increasing cover, then the LSA is in line with the climatic model; but if one looks at the medium-size antelopes, then the picture seems to be contrary to the overall climatic picture. This apparent dilemma is cleared up somewhat by a consideration of Relationship A in Figure 5.11. This line traces the relationship between eland MNIs versus the small, cover-loving antelope (grysbok and bush-buck). This line should monitor the relationship between hunting versus scavenging, with Relationship A representing scavenging. What is most interesting is that Relationship A and Relationship B are very strongly correlated. This means that the exploitation of cover-loving, medium-size animals is part of the scavenging strategy, whereas the exploitation of the grassland antelope occurs primarily as a component of the hunting strategy in the MSA. Put another way, the grass-loving animals—hartebeest, wildebeest, and bastard hartebeest—seem to vary with whatever is conditioning the relative roles of hunting versus scavenging, and do not necessarily betray directly environmental conditions. If this is correct, then the grass-loving forms should be primarily represented by young individuals because the hunting bias seems to be clearly in favor of small prey, whereas the cover-loving antelope seem to vary in terms of scavenging biases, which, as we have seen, favor large-body-size animals. Presumably, then, the cover-loving antelope should be primarily represented by adults, and the grass-loving forms should be primarily represented by young, immature animals. The apparent trend toward decreasing cover-loving medium-size animals, which at first glance seemed to parallel expectations for a deteriorating or increasingly cold environment, is now recognizable as just another manifestation of the trend toward increased hunting and a more and more marginal role for scavenging throughout the Klasies Cave 1 sequence.

This insight brings us back to a reconsideration of the seeming anomaly of the LSA. I think that most archaeologists would accept as fact the statement that the people responsible for the LSA were hunters equipped with a technology that was capable of effectively taking moderate- to large-body-size animals. The ethnohistoric data, although acknowledging the role of scavenging (see Figure 5.8), clearly depict it as a very minor and expedient set of tactics. Active taking of prey either through hunting or trapping dominated the strategies for obtaining animal products among the relatively recent occupants of the region. It is therefore suggested that the relative proportions between medium-body-size animals versus small-body-size animals in the LSA assemblage cannot derive from the same tactical causes as their proportions in the MSA assemblage.

The proportions between medium to small body-size in the LSA may reflect the relative roles of trapping versus hunting, and/or a density-dependent reduction in landscape available for noncompetative hunting. After all, domestic animals were adopted and husbandry was present in the Southern Cape by around 2000 B.C. This had to solve some local problem.

It is likely that, during the terminal Pleistocene, population built up as it did in other regions, and mobility was increasingly restricted. This forced an intensification with the regular use of smaller and smaller food packages (see Binford 1983b:195–213 for this argument). We can view the increased exploitation of small antelope as well as marine resources not as a function of increased hunting, as in the MSA situation, but as a response to more and more circumscribed ranges and the related intensification in the use of more and more localized resources. Such resources are necessarily smaller. This would mean that when moderate- to large-body-size forms were taken, they would more likely be migratory and "unearned" relative to the local habitat, or obtained through expedition hunting. These would, of course, be the nonlocal, migratory, grassland feeders, which in either case would occasionally enter the Cape province, perhaps during dry conditions from the *karoo* and the grasslands to the north, or would be introduced by logistically organized expeditions into that area.

In any event, the relative frequencies of similar species can be expected to exhibit marked variability between the LSA and the MSA, reflective of differences in the organization of subsistence strategies. The degree that species-frequency data also reflect differences in environments is something that cannot be read directly but must be evaluated by seeking to control for different variables with surrogate measures, as I have tried to do in this discussion.

The conclusion is clear: the patterns for both MSA and LSA in relative species frequencies are dominantly reflective of hominid ecology and not gross environmental conditions per se. This conclusion has major implications for both the comparative chronology of the MSA and LSA but also impacts the current attempts to view comparative faunal frequencies between very different ecotypes as surrogate, direct measures of environmental dynamics (see Beaumont 1980; Klein 1980; Volman 1981). We do not know how the hominids were exploiting the grassland fauna in a grassland setting. How does one compare the biased taking of young, with little or no scavenging of such animals as wildebeest and bastard hartebeest, as seen in the Cape setting, with exploitative tactics in the grassvelt? We simply do not know. Granting that variability among species as a direct environmental measure is inappropriate, all the currently constructed chronologies based on faunal–environmental equations for the MSA are strongly suspect. Environmental inferences can only be facilitated through an intervening understanding of the strategies and tactics carried out by the hominids who made use of the resources.

Dating of the Site Sequence

The faunal analysis has permitted the recognition of some very interesting trends. These trends are believed to be behaviorally directional and perhaps irreversible, at least in this region. The major trend was a steady increase in the hunting of small-body-size animals throughout the MSA sequences represented at Cave 1 (the exception being Level 13). There was a concomitant decrease in scavenging the carcasses of medium- to large-body-size animals. There appears to be a bias in favor of scavenging the local medium-size antelope, while more grass-loving animals of the same size seemed to have been more commonly hunted, presumably as young and juvenile individuals. The trend toward increased hunting ensured an apparent increase in the frequency of grass-loving forms throughout the sequence. Accepting this trend as general, and not site-specific, permits us to offer a plausible relative dating for the stratigraphically discontinuous deposits excavated from Shelter 1A and 1B at Klasies River Mouth. The values for the various percentages of the noncontiguous levels are given in Table 5.6. If one fits these values to the curves shown in Figures 5.09–5.11, it becomes clear that the so-called MSA II from Shelter 1A is best accommodated between Levels 14 and 13 of the Cave 1 sequences. Similarly, the Howieson's Poort levels of Shelter 1A may well have been accumulated during the maximum low-water period represented by Level 13 in the Cave 1 sequence, whereas MSA III of Shelter 1A probably postdates the deposits of Cave 1 and may represent a period of warming temperatures after a period of maximum cold and hence low sea-level.

Of particular importance to the arguments regarding the types of humans responsible for the depositions at Klasies River Mouth is the relative dating of the deposits from Shelter 1B. Comparison of the values for the various species percentages given in Table 5.8 clearly indicates that it must date during a period of substantial hunting and relatively little scavenging—which is, as can be seen in Figure 5.9, after Level 16 in the Cave 1 sequence. Given the very low figure for the eland/small-bush-cover antelope comparison (Table 5.8, Row 26), it is most likely that the deposits postdate the Level

13 deposits, placing it contemporary not with MSA I of Cave 1, but with MSA III of Shelter 1A. Based on the faunal seriation developed here, the chronological sequences for the various depositional zones at the Klasies sites would be from oldest to youngest: Cave 1 Levels 39, 38, 37, 17b, 17a, 16, 15, Shelter 1A MSA II (Levels 22–23), Cave 1 Level 13, Shelter 1A Howieson's Poort (Levels 10–21), fill of Shelter 1B, and finally MSA II from Shelter 1A (Levels 1–9). Obviously there may be some overlap among the sets of levels.

This reinterpretation of the chronology at Klasies River Mouth removes one of the very puzzling implications of the sites; namely, the belief by the original excavators in a very early presence of anatomically fully modern man, and in the contemporary presence of at least two types of humans—"Neanderthaloid" robust form and the more gracile "fully modern man." This picture was created for the excavators by their equation of Shelter 1B with the lower levels of Cave 1. The faunal seriation suggests that actually the contents of Shelter 1B are partially contemporary with or later than the Howieson's Poort levels of Shelter 1B. The presence of a gracile form approaching fully modern man at this place in the sequence is not very surprising. Removing this example from the seeming mixture of robust and gracile remains claimed for the Klasies site leaves only the remains referable to Level 14 (Cave 1), which is almost certainly derived from scree, and is at least partially a secondary deposit contemporary with, or just slightly younger (on average) than, the Howieson's Poort levels of Shelter 1A. Having modern forms of humans associated with levels that yield evidence for personal ornamentation is not surprising.

Thus far I have avoided mentioning any estimates as to the absolute age of the deposits. As was pointed out earlier (Chapter 2), the only data cited to justify the dating of the lower levels of Cave 1 to the isotopic Stage 5d–5e dating approximately 125,000 B.P. rests with the comparative data on $^{16}O/^{18}O$ ratios measured for six shells—two from the LSA deposits and four from the MSA I. Both the data base and the inferences as to the meaning of similarities provide a most tenuous basis for dating. If we dismiss the comparisons between LSA and MSA I $^{16}O/^{18}O$ ratios, we have no basis for rendering an estimate of absolute age. There is simply no factual basis for any positive claims for great antiquity of the gracile human remains nor of the industrial facies called Howieson's Poort beyond the fact that they apparently date beyond the reliable range of ^{14}C methods. This means that they are essentially older than 35,000 years, which does not demand any extraordinary role for the South African region in the history of the appearance of fully modern man. The data indicate increased hunting, technological changes that may well be related to hunting technology, and the appearance of gracile, fully modern human forms. At present these are best considered to be roughly contemporary with analogous changes in other parts of the world.

Summary

It has been argued that throughout the MSA sequence recorded at Klasies River Mouth, particularly in Cave 1, there is a trend toward increased hunting of small game and the young of large species, with a corresponding decrease in the dependence upon scavenged foods obtained generally from larger-body-size animals. This pattern is unrelenting through the Cave 1 sequences, except for the contents of Level 13, which is believed to have accumulated during a low-water stage and is reported to contain no obvious hearths nor lenses that can be attributed to occupational use of the surface during accumulation.

It is unclear at present whether the low frequency of small animals (as measured by grysbok and bushbuck) is referable to preservational bias or sorting in secondary deposits, or is an accurate behavioral indication. The very small sample size cannot be overlooked in assessing the meaning of the deposits' contents. All in all, the pattern as manifest suggests a major temporal trend that may be shown to be partially sensitive to environmental changes or topographic setting. The demonstration that there is a strong behavioral shift in the relative roles of hunting versus scavenging provides the basis for significant shifts in the relative frequencies of various species. These shifts have in the past been read as a direct reflection of environmental change. This analysis challenges such methodological conventions, while at the same time leaves open the possibility that the temporal patterning noted between hunting and scavenging may be at least partially responsive to environmental conditions. In many cases of culture change, a practice that has been primary may take on more specialized and restricted roles in the overall organization of the system as its primary role is replaced by a more effective set of alternative tactics. This is an area in need of investigation and represents a domain of our ignorance unappreciated as long as relative species' frequencies are conventionally interpreted as simply reflecting climatic change.

Because the correlation of climatic episodes with an overall pattern of climatic fluctuation had been the major method of dating these deposits, the implication of the species' frequencies to adaptive change within the hominid niche renders the inferred chronology of the deposits and the alleged evidence for very early, fully modern man in the southern African setting strongly suspect, if not totally obselete.

What Have We Learned from Klasies?

I suppose the most important argument to come from this research is that there does seem to have been a directional trend evidenced at Klasies River Mouth in the relative roles of scavenging versus hunting. Hunting seems to have increased regularly through the major part of the sequence at Cave 1 in its contribution to the diet of the hominids. This places the view of the MSA in a dynamic mode. For instance, the earlier studies of the MSA noted that there appeared to be considerable interassemblage variability and even that there was an alternation of industries, but there was no clue to what this variability might reflect.

Is the apparent interruption in the hunting trend seemingly indicated during the low-water conditions suggested for Level 13 in Cave 1 and the Howieson's Poort of Shelter 1A a clue? I have noted previously that Middle Paleolithic assemblages in European sites tend to remain relatively stable over long accumulatory periods and then to change coincidentally with evidence of major environmental change. Although there is a coincidence to such changes, there is rarely any correlation between the type of environment (see Binford 1982a). I have always thought this represented spatial repositioning of the land-use pattern in response to environmental changes, but commonly with little change in the organization of the adaptation *per se*. The alternation of industries as well as the break in the hunting trend seen in the Klasies sequence may well represent an analogous situation, with the added possibility that the old repositioning strategy was still going on coincident with a major trend in adaptive reorganization and increased reliance on hunted foods.

The impression that many of us have had, that the site-use patterns of the Middle Paleolithic were fundamentally different from those of the Upper Paleolithic may be enlightened by the suggestions made here that sites like Klasies were not *base camps* in the traditional sense of the term. They may have been functionally specific places. The midday waterhole site and the nighttime protected sleeping site may have been just that—special-use areas, which were not provisioned like base camps. Much feeding could well have taken place at the points of food procurement. Only if special processing prior to consumption was needed—such as soaking, breaking open, and cooking—would food be transported to appropriate processing places. In addition, there may have been an attempt to extend the feeding potential of resources by transporting some into protected locations. Clearly the food packages introduced to Klasies River Mouth were small. It is hard to imagine them as the basis for regular food sharing and the provisioning of a camp in the sense characteristic of modern man.

At Klasies River Mouth, the hominids also fed along the beach. As in the case of terrestrial foods, they introduced small food units, seemingly discrete shells, into the site. Part of the uneasiness that many of us have felt when seeing the typical pattern of a fully modern man living in base camps far back into the Pleistocene may be partially relieved by the suggestion that Klasies was a nighttime sleeping site, and a midday water source location is implied by the scavenged parts from animals likely to have been killed by nocturnal predators near a water source. Remains from scavenging may appear dominant in the sleeping site simply because they required processing—transport being not for provisioning a group but simply an accommodation to the localization of processing facilities, tools, and protection. It appears to me that we have the opportunity to come to considerably more interesting conclusions about assemblage variability in the MSA than that it simply represents "changing fashions" (Volman 1981:259).

In the past there was no clue to dynamic trends that might have been going on among the hominid populations immediately ancestral to the fully modern humans that appeared in southern Africa near the end of the MSA. I think it is clear that hunting was increasing, at least in the southern Cape province. Many of the changes in technology that typify the MSA, such as the manufacture of bifacial and unifacial points, as well as the appearance of crescents or backed knives, betray some specialization in tool production and innovations in tool design. These must be related to some tool-use demands. New demands can be expected to be associated with new behavioral trends and shifts in adaptive tactics. This places the South African data squarely in the middle of our resources regarding the fascinating transition from Middle to Upper Paleolithic, considered in behavioral terms (see Binford 1982a; Mellars 1982; White 1982). Although some of us have noted major contrasts between the archaeology of the Middle Paleolithic, as it is termed in Europe, and the Upper Paleolithic, none of us ever considered that one major basis for some of the contrasts might be that during the Middle Paleolithic hunting played a much-reduced role relative to the adaptations of the Upper Paleolithic. In short, although much attention has been placed on the problem of the transition in the Northern Hemisphere materials, we never had any clear-cut trends that were thought to provide the dynamic context in terms of which evolution proceeded. I am convinced that the Klasies data is supplying us with just such a trend in the Southern Hemisphere.

Perhaps the most surprising result of this analysis is the seemingly important and consistent role of scavenging among hominids who are, by all Pleistocene standards, very late indeed. What does this imply about adaptive strategies characteristic of earlier African hominids? Is this a biased view by virtue of seasonal occupation at Klasies River Mouth? Does the little section of temperate environment at the southern end of Africa provide us with an important control on arguments as to the ecological pressures guiding hominid physical and behavioral evolution as hominid populations successfully radiated out of more tropical settings?

CHAPTER 6

Beyond Klasies River Mouth: Implications for Understanding Early Man

In the introduction I pointed out how ideas regarding both the context of evolutionary change and the history of hominid evolution have been phrased in scenarios of what was termed *evolutionary functionalism.* Most often these arguments were based neither on a knowledge of the sequence of changes that took place nor on any in-depth understanding of the evolutionary mechanisms operating to bring about changes. In recent years, we have begun to develop a kind of chronology of changes. We know a great deal more about the history of bipedalism, and we need to speculate much less about the actual history of changes in brain size and facial structure within the hominid line. Unfortunately, however, the thrust of much archaeological research has not been in the direction of developing reliable diagnostic methods for recognizing behavioral characteristics that many have thought important in the context of changes leading to our modern condition. I have commented earlier on the archaeologist's tendency to build models and then to argue that the data from the past can be accommodated to whichever model the archaeologist prefers. Certainly with such a methodology we will never learn what the past was like. Instead, we only learn how archaeologists can invent accommodating arguments or make a priori assumptions about what the past was like. We need, instead, reliable methods for decoding the archaeological record to obtain an accurate glimpse of the past.

Is Klasies River Mouth Unique?

Before considering the implications of the Klasies study, I think it is important to address the issue of how general the types of inferences drawn from the Klasies data might be. If Klasies River Mouth Cave 1 is a totally unique site or represents a totally unique adaptation, then there are essentially no implications behind the recognition of variability in the behavior of near-modern man. This is a point that should be appreciated, but certainly not one to excite a great deal of discussion. On the other hand, if the analysis presented here has merit and if Klasies is not unique, but indeed is informative about more general behavioral conditions at the time, there are certainly important implications for our current thinking about early man.

The tactics summarized here, of scavenging coupled with opportunistic killing of small animals, were the basic carnivouous tactics of hominids living at Klasies as recently as sometime just before 35,000 to 40,000 years ago. This is the period of time contemporary with the Mousterian of Europe and the Near East. It is removed from the lifeways of the hominids of the Plio–Pleistocene boundary by approximately 1,500,000–1,800,000 years! I have not been describing the archaeological remains of some "dawn man," practicing a way of life at the very beginnings of our ability to recognize human characteristics. I have been describing the faunal remains from hominids who have, in fact, been suggested to be anatomically indistinguishable from both fully modern man and their immediate predecessors (see Beaumont 1980; Deacon 1981; and Rightmire 1979). These were clearly hominids who should have been enjoying the fruits of our mid-Pleistocene evolutionary progress:

> the dry conditions determined the adoption of a partially terrestrial life and the eating of meat, first as parts of a mixed diet. *The Australopithecines and such types as KNM 1470 represent this stage,* whether or not they are on the direct line of human ancestry. *Then from about two million years ago, with rapidly varying and often severe conditions came the changes in the brain and endocrines by which the hominids became social hunters, such as* Homo erectus, *using implements, sharing the products of the quest for food and developing a simple culture.* The advantages conferred led to still more rapid increase of brain power, which allowed for survival through the often harsh conditions of the later Pleistocene. Finally further developments in the same directions made possible the logic, language, and culture that marked the emergence of man as we know him when conditions became milder in the Neothermal period. (Young 1981:215–216; emphasis added)

The above view of human evolution was synthesized after a recent meeting, in which a large number of scientists presented what they considered the most up-to-date data and interpretative arguments available bearing on the problem of how we came to be. If the above view is even close to being accurate, the subsistence behavior suggested here for the occupants of

Klasies River Mouth should have been extinct at least since the appearance on the evolutionary stage of *Homo erectus,* the "social hunter."

Perhaps the Klasies River occupants were unique and represented only a local variant (a conservative group like the early evolutionists considered the Tasmanians to have been), or even a throwback to forms of behavior that were more general at a much earlier age!

I think an attempt to dismiss the Klasies case as abnormal can be dealt with in a variety of ways. Perhaps most important is the patterning that Richard Klein has noted as characteristic of other MSA sites. Klein (1975b) has generalized that the large-animal "head and lower legs" pattern of anatomical-part frequencies is characteristic of early living sites, although at the same time the small animals are represented by more meat-yielding parts, as at Klasies River Mouth.

In speaking of the important site of Border Cave, Klein states: "with regard to the large bovids . . . they are represented almost entirely by parts of the skull and feet; their limb bones and vertebrae are very rare. . . . Suids and zebra are also characterized by patterns of body part representation in which foot and skull bones predominate heavily" (Klein 1977b:24). Other important southern African sites exhibit analogous patterns (see Thackeray 1979).

Judging from the literature, the pattern of head and lower legs from large animals and more complete anatomical representation for the smaller animals seems to be very general at sites in southern Africa during the MSA.

The analysis developed here is also surprisingly consistent with earlier analyses and the facts as they are increasingly appreciated from the early hominid sites at Olduvai Gorge. The hominids at the Olduvai sites were probably scavenging the same range of anatomical parts as indicated at Klasies River Mouth Cave 1; the only difference is they were processing them "in the field"—the midday-rest location near a water source. The previous analysis of the Olduvai sites suggested that there was a strong component recognizable as normal killsite assemblages characteristic of nonhominid predator kills (see Binford 1981:273–288). The presence of a carnivore-killed bone assemblage is consistent with the midday-rest locations discussed here.

Similarly, the analysis that I carried out on the Olduvai sites indicated that there was a residual pattern of covarying bones, which were tentatively identified as related to hominid behavior. The bones so identified were recognized as basically those that "yield only bone marrow as edible material" (Binford 1981:291). Again, this is perfectly consistent with the scavenger patterns identified at Klasies River Mouth, and are the parts that should regularly appear at a midday-rest site. There are many other characteristics, such as the reported high frequency of hyaena-damaged bones (see

Potts 1982; Potts and Shipman 1981), which in terms of the descriptions of species dynamics around waterholes should be present in substantial quantities, given the bone-concentrating behavior of hyaena around water sources (see Binford 1983b:62–70). Finally, the observation of a high frequency of cut marks on metapodials at Olduvai is completely consistent with the patterns described here (see Lewin 1981).

The consistency between the Olduvai data and the information from Klasies strongly supports the view that Klasies is not unique, but additionally implies that a long-term pattern of very consistent behavior links Klasies and the very ancient past.

Although the analysis of ancient European and Asian faunas is complicated by the role of carnivores as contributors to the populations of bones commonly attributed to man (see Binford 1981 for discussion of this problem), there are some provocative patterns nontheless. For instance, the famous site of Lazaret is characterized by a head-and-lower-leg pattern for the large- to moderate-size ungulates, whereas the most-common species at the site is rabbit (see De Lumley 1969). The seashore setting of Klasies River Mouth demands some comment relative to the early seaside site of Terra Amata, near Nice in France. Like Klasies River Mouth, it was near the sea, and like Klasies there is some evidence for the use of aquatic resources. Detailed faunal data are not available to me, yet it is certainly provocative to repeat De Lumley's comment on the fauna:

> Although the visitors did not ignore small game such as rabbits, and rodents, the majority of the bones represent larger animals. These are in order of the abundance, the stag (*Cervus elephas*), the extinct elephant (*Elephas meridonalis*), the wild boar (*Sus scrofa*), the Ibex (*Capra ibex*), Merk's rhinoceros (*Dicerothinus merki*) and finally the wild ox (*Bos primigenius*). Although the hunters showed a preference for big game, *they generally selected as prey not the adults but the young of each species.* (De Lumley 1969:49; emphasis added)

One can only wonder what the relative frequencies of anatomical parts across an individual size gradient would look like. There are still further facts that render European faunas consistent with many of the arguments presented here, of which the disputed characteristics reported from the Spanish site of Cueva Morin (see Altuna 1971:392; Binford 1983c; Freeman 1983) provide a further case in point.

Finally, the data from Klasies River Mouth are not anomalous relative to a number of generalizations that have been previously offered regarding long-term trends in faunal utilization by African hominids. However, if one believes that early man was an effective predator, then the facts of the archaeological record must be accommodated in novel and interesting ways. For instance, Garth Sampson, speaking of the Acheulian sites of southern Africa, comments as follows:

> The evidence suggests that large game animals provided much of the Acheulian meat supply.
>
> Whereas the bones of the lesser game could be obtained by scavenging from carnivore kills, the presence of the very large animals (presumed to be beyond the hunting capability of carnivores) must reflect organized hunting and probably trapping by man. There is no evidence for the use of pit traps in Acheulian times, but it has been argued (Clark 1959) that they must have been used to kill the larger animals. (Sampson 1974:128)

Sampson, in discussing the shifts in faunal remains indicated by comparing the later MSA sites with the Acheulian sites, summarizes the situation as follows:

> The available fossil evidence from Acheulian sites hints at a marked selective bias in hunting activities toward the larger animals. . . . The total absence of elephant and the extreme scarcity of rhino and hippo from Pietersburg and Bambata sites may even suggest a shift in hunting goals in the Post Acheulian period. Indeed, the range of animals taken at this time would suggest a more random selection of any available foodstuffs and a more systematic exploitation of different microhabitats surrounding these sites. (Sampson 1974:215–216)

Continuing his comparisons, Sampson further notes increases in the frequencies of smaller and smaller species, commenting that on the assumption that the hunters took a random sample of the local game population, "a decrease in the number of large species in the population would, therefore, be reflected in the hunter's sample" (Sampson 1974:246). Tracing the trends in faunal remains into still later time, Sampson notes that in the Wilton and Oakhurst complexes, "intensive exploitation of all local food resources including a wider range of small animals from different niches" is indicated. He further notes: "Large game animals are not abundant at the listed sites" (Sampson 1974:398).

In Sampson's work we see the fascinating situation of his postulating traps and other devices to account for the assumed early hunters' bias in favor of very large animals, and then postulating environmental change as the cause of early man's seemingly increasing exploitation of smaller and smaller species as we approach the recent past. The pattern seems clear: the very early sites (largely waterside sites) have large animal remains, and in the MSA we begin to pick up increasing numbers of cave occupations in which small species tend to dominate, and in the Klasies data the large species are represented by head-and-lower-leg-part profiles. Finally, with the appearance of behaviorally modern man, whom we know to have been hunting, moderate- to small-body-size prey seem to dominate.

The analysis of the Klasies data leads us to infer very different meanings for the demonstrable pattern of decreasing hominid association with large animals, and increasing association of more recent men with increasing numbers of small- to moderate-body-size animals as described by Sampson.

Very early man is probably most commonly represented by his midday-rest sites or feeding locations, where scavenging took place. Only with the increasing number of cave occupations in the MSA do we begin to get a glimpse of the hominids' sleeping locations. This shift roughly corresponds to an apparent initial increase in the taking of small- to moderate-body-size prey, as discussed here for Klasies River Mouth. Finally, with behaviorally modern man, scavenging becomes a rare subsistence tactic, whereas hunting assumes more importance and is therefore reflected in more moderate-body-size forms.

Implications for Our Ideas of the Past

If we accept Klasies as representative of an era, the implications of this study are several: (1) I have sought to develop a set of methods for inference that permit the recognition of hunting versus scavenging tactics; (2) I have discussed the historical implications of the use of this nascent methodology on the data from the site of Klasies River Mouth; and (3) the inferences drawn from the above two phases of this study reflect a number of key issues regarding the current state of paleoanthropological inferences about life in the ancient past, and substantially impact a number of ideas about the past that may be misguiding the way evolutionists approach the problem of understanding our heritage. In light of the diverse nature of the implications, I break this final chapter into a number of sections, each treating a slightly different implication that I consider important.

Hunting: the Rise of Intelligence and Other Food-Related Ideas of Evolutionary Causation

I suggested in the introduction that a common idea linked the rise of large-brained hominids to a shift to a hunting subsistence strategy. This shift was commonly considered to have been forced on the hominids by their environmentally induced entrance into savanna–grassland settings. Hunting was thought to favor cooperative male behavior and increased intelligence. In very simple terms, a shift in food-getting strategy to a predatory set of tactics favored intelligence and a changed sociality.

The ideas about subsistence strategy that I have attempted to warrent as relevant to the past are neither new nor novel. As I pointed out in the introductory discussions, the idea of scavenging was advanced almost as early as there was knowledge about our early hominid ancestors. Perhaps the first ecologically informed suggestions were set forth by George Schaller, who was then deeply involved in the investigation of predators. Schaller summed up his suspicions nicely as follows:

> Like all predators, hominids probably obtained their meat in the easiest possible way, by scavenging and by killing the young and sick when possible, but pursuing healthy animals when nothing else was available. . . . The scavenging and hunting hominids' primate heritage suggests that they were diurnal. . . . [A]n ecological opening existed for a social predator that hunted large animals and scavenged during the day, an opening that an early hominids may well have filled. (Schaller 1972a:68; with permission from *Natural History,* Vol. 81, No. 3; Copyright the American Museum of Natural History, 1972.)

Although Schaller was very close to recognizing the particular subsistence role that early hominids might have played in the African homeland, even he seems to have been giving the early hominids credit for taking mature large game. This analysis suggests that this activity appears quite late in hominid behavioral evolution.

As in so many cases, I have found the writings of J. Desmond Clark to have an almost uncanny, prophetic character. He has for many years anticipated the research directions needed to move our understanding of early man away from pure speculation (see Clark 1965). In the area of subsistence research he recently wrote:

> Studies of food waste on sites of proto-hominid activity suggest that much of the meat was consumed on the "living sites" was obtained by scavenging (Vrba 1975). For the early hominids to be able to compete successfully with carnivores, it is possible that the particular adaptive niche they occupied was that of "middle-of-the-day scavengers" (Schaller 1972). (Clark 1980:43)

Obviously, the idea of scavenging playing an important role in hominid subsistence and the exploitation of young and small animals has been previously considered, but generally in the context of a gradualist view of evolution, in which these strategies were expected to be transitional between a basic plant-based diet and one more dependent upon large animal predation.

What is new with this study is the recognition that what was thought to be a transitional strategy, perhaps characteristic of the early hominids at the Plio–Pleistocene boundary, was the regular strategy of a group of hominids living perhaps as late as 40,000 years ago. Certainly any arguments that would seek to make hunting in the big-game sense of the word a major molding force for explaining increases in brain size, or the morphophysiological shaping of modern man are therefore suspect.

Hunting Reconsidered

The trends documented at Klasies are provocative in a number of ways: (1) the site is in a temperate zone; (2) regardless of the controversy over dating, it is certainly late in the Pleistocene; and (3) there is a demonstrable trend toward increased hunting and decreased scavenging. This trend appears to be coincident with the appearance of fully modern man in the bioanatomical sense of the term. As in other places, the appearance of fully modern man is coincident with the appearance of items of personal ornamentation and other evidence of symbolic behavior. Most earlier arguments have tried to make hunting the behavioral context in which selection operated to bring into being our humanness—in the modern sense of the word. The Klasies data stands as a caution: Is it possible that hunting and all that it implies in terms of planning (see Binford 1982a) may well be a part of the emergence of our humanness in a modern sense?

Meat eating and the eclectic feeding on animals and animal products as edible products encountered in the habitat may well have been relatively early and an important preconditioner for hunting as an organized strategy. Nevertheless, omnivorous diet does not make one a predatory strategist. Just as relative species' frequencies may mean something very different in the LSA than the same species frequencies do for the MSA, the presence of animal parts and products in sites of early man may well indicate very different forms of organized behavior than have yet been imagined for our ancestors.

Tools, Weapons, and Hunting Aids

I have stressed the facts that the faunal remains were consistent with the idea that animals being exploited for their meat were small; individuals whose body size was less than 90 pounds (41 kg), with an average that appears to have been considerably less. In addition, the individuals taken were the young of ungulates who largely practiced a "hider" strategy, or adults who were solitary, territorial creatures, most active nocturnally. This means that human color vision, operating in the daytime, was a distinct advantage. It located stationary, camouflaged prey that may well have been simply overpowered by a hominid killer. No evidence for "killing at a distance," as with spears or more complicated projectiles, seems indicated. Similarly, other capturing tactics, such as traps and surrounds, do not seem consistent with the selective taking of isolated camouflaged prey. The idea that tools were invented in the context of hunting seems unlikely. They were most likely invented to solve processing and procurement problems, such as the removal of a dried limb from a desiccated carcass. They must be seen as technical aides in overcoming problems in feeding, and not as some breakthrough making possible new characteristic behaviors. To be sure, that happened later, but I strongly doubt such roles for the earliest tools.

Home Bases and the Altruistic Sharing Model of Human Evolution

In recent years, certainly the dominant model of early hominid life has been the home-base sharing model popularized by Glynn Isaac (1978a, b) and adopted in most of the recent popularist literature (R. Leakey 1981). The analysis of the Klasies River Mouth data provides strong support for the inference of a type of site, located adjacent to a water source, that would in no sense be a home base. This is of course the midday-rest location, already discussed, where the hominids would have had access to carcasses previously ravaged by nocturnal predators—animals that were essentially dust-to-dawn drinkers and feeders. The famous sites of Olduvai Gorge immediately present themselves as possible examples of such locations. My earlier analysis of the Olduvai sites suggested that some of the patterned, bone-frequency configurations at those sites were indistinguishable from carnivore kill sites (Binford 1981:281–282). The Klasies data suggest that hominids at such water-source locations would be scavenging an association of species that Ayeni's data identify as those most likely to be killed by large nocturnal predators around a water source.

In a similar fashion, my analysis of Klasies River Mouth leads to the conclusion that hominids were scavenging marginal foods from previously ravaged carcasses. The data from Olduvai Gorge indicate unequivocally that hyaenas ravaged many of the bones remaining on the Olduvai sites (see Potts 1982; Potts and Shipman 1981). The data initially reported from Olduvai are also consistent with the data reported from Klasies River Mouth—many of the tool marks were inflicted on metapodials and otherwise non-meat-yielding parts. This observation is consistent with my observations on the "unexplained" high frequency of lower-limb bones on the Olduvai sites (see Binford 1981:278–288, particularly p. 281). It seems likely that most of the Olduvai sites represent the accumulation of bones deriving from overlapping use of the same spots by lions, hyaena, hominids (see Figure 6.1), and ungulates.

The convention that has been used over and over again, that a co-occurence of stone tools and high bone-densities betrays a home base (see Isaac 1971:282–287), no longer seems tenable. A midday-rest location, where bones scavenged from the carcasses of other carnivore's prey were carried short distances for processing as marrow bones, results in an association of stone tools and utilized bones, but this is almost certainly not a home base. This argument has a major set of implications for the inferential methods employed by many archaeologists. It is clear that if my argument regarding midday-rest locations can be sustained, then the convention that a functional association between stone tools and animal bones can be taken as indicative of a home base is clearly rendered ambiguous and, as such, unre-

Figure 6.1 Reconstruction of hominids processing lower-limb parts at a water source.

liable as a justification for inference. This argument also implicates a number of other research strategies and/or argumentative tactics currently being explored by paleolithic archaeologists. For instance, Isaac (1983b) has advocated a view of the early sites as "dynamic through-flow systems" (see Figure 6.2).

Although I certainly acknowledge that many agents may contribute to or modify the associations generated by hominids at points in their environment, this perspective never comes to grips with a basic problem: How we recognize a base camp? Isaac seems to consider all the alternatives as arguments of relevance (Binford 1983a:175–161), in which some interpretations might bring into question the applicability of the home-base model to a particular site, but they never actually bring into question nor offer an alternative to the home-base behavioral model of organized hominid land-use. For instance, Isaac has recently summarized what he considers to be multiple working hypotheses regarding the interpretation of demonstrable associations of stone tools and bones. Figure 6.3 summarizes these alternative. The first alternative, the "hydraulic jumble," is certainly a possibility and could well be relevant to some sites, but establishing this would not negatively impact the idea that others were home bases if they could be shown not to be hydraulic jumbles. Similarly, his second argument, where a "common amenity" is assumed to account for evidence of hominid actions and carnivores, in no way negates the idea of a home base. The third alternative, the use of "carnivore accumulations" by hominids, is simply another form of the second alternative, in which once again man could be using the place as a home base after carnivores had completed their ac-

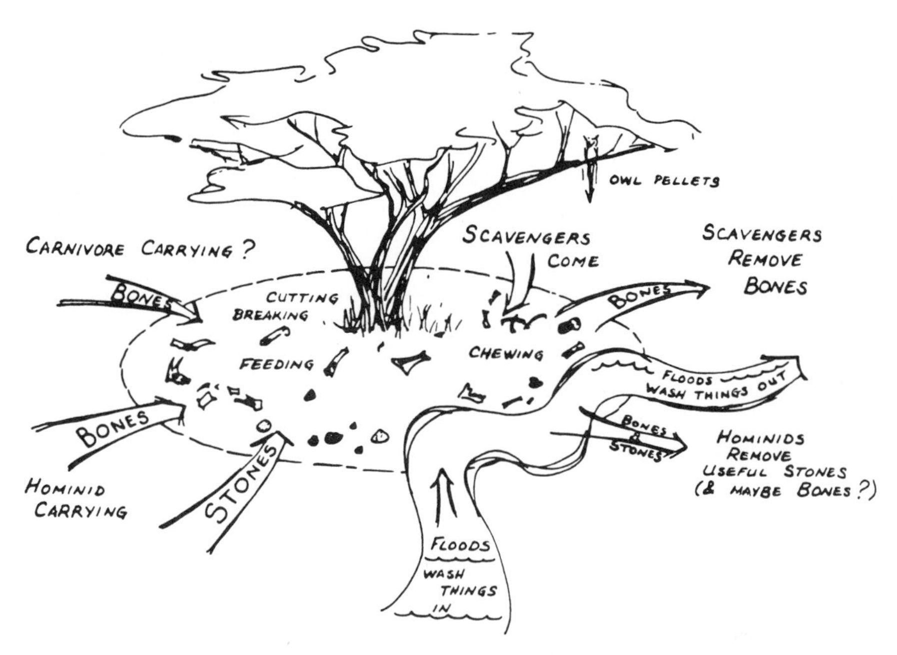

Figure 6.2 Glynn Isaac's (1983b) "dynamic flow through model" of site formation.

tivities. In similar fashion, the fifth alternative, "central-place foraging," argues that if hominids positioned home bases where food opportunities existed, then this in no way precludes the organized use of a home base. All Isaac's "alternatives" appear as various event sequences that could account either for the placement of a home base or for the association between home-base activities and the behavior of other animals.

The implications of this study discussed thus far could all be challenged by convincing demonstration that the interpretations of the Klasies data are inappropriate. Similarly, the implications about hunting, scavenging, and even alternative forms of hominid behavior leading to associations of stone tools and hominid-modified bones are subject ot debate, but it seems to me the methodological implications of my discussion for Isaac's use of the home-base model are not contingent upon the Klasies analysis per se. Here we face a logical problem of research tactics and a problem of research strategy. Isaac cites modern systems as his justification for believing that functional associations between stone tools and bones mean home bases—they generally do in modern systems. Given Isaac's justification, the citation of the association is considered prima facie evidence that the modern bome-base form of the system was responsible for the association between stone tools and animal bones observed archaeologically. In spite of Isaac's frequent appeal to the method of "multiple working hypotheses," he has no

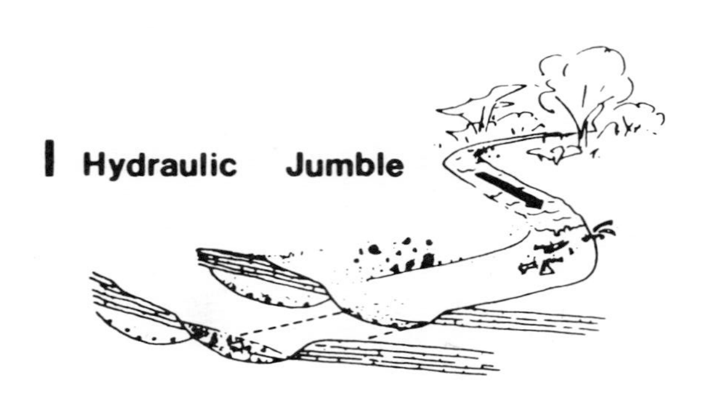

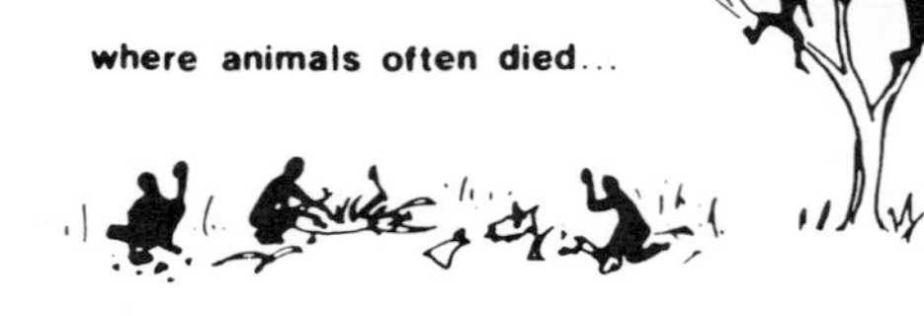

Figure 6.3 Glynn Isaac's (1983b) scheme of alternative interpretations for early site formation.

alternative model regarding the way early man was organized adaptively. This means that Isaac's argument has the form of a logical tautology in which the conclusion is given by the premise, and no evaluation of the premise is attempted. In short, the world of the ancient past is accommodated to his belief about organizational form by virtue of an interpretative convention. In the development of archaeological methods, it is the validity of the premises from which we start that is crucial. If, as in the case of Isaac's approach, these premises are suspect, then of course all statements reasoned from these premises are suspect.

I have offered a set of possible hominid behaviors that could yield regular associations between stone tools and animal bones, yet not imply sharing, home bases, and so forth. I have raised the specter of ambiguity over the interpretative conventions commonly in use. At this point we need to explore the possibility that still further ambiguity may surround the facts in dispute—associations of stone tools and animal bones—but at the same time seek ways to reduce the ambiguity and therefore provide diagnostic criteria for recognizing one possible condition from another. As Gould has so cogently said:

> it suggests a false concept of how science develops. In this view, any science begins in the nothingness of ignorance and moves toward truth by gathering more and more information, constructing theories as facts accumulate. In such a world, debunking would be primarily negative, for it would only shuck some rotten apples from the barrel of accumulating knowledge. But the barrel of theory is always full: sciences work with elaborated contexts for explaining facts from the very onset. Creationist's biology was dead wrong about the origin of species, but Cuvier's brand of creationism was not an emptier or less developed world view than Darwin's. Science advances primarily by replacement, not by addition. If the barrel is always full, then the rotten apples must be discarded before better ones can be added. (Gould 1981:321–322)

In recognition that it is only through the recognition of alternatives that we even begin to suspect what may be a rotten apple, I have attempted to think through an alternative model of hominid land-use, which then may provide the comparative framework for coming to a judgment as to whether one should replace another in the already full barrel of method being used daily in the interpretation of early man.

An Alternative to the Central-Place Foraging Model of Hominid Behavior

Advocates of the sharing hypothesis universally make the assumption that the motive for transporting items, particularly potential foods, is in order to share with others. In fact, this assumption leads Glynn Isaac and his co-workers to act as if the home-base interpretation becomes self-evident once an association of stone tools and animal bones is demonstrated. It only seemed plausible to imagine the transport of food by the finders to the place where other consumers could share in the finders' good fortune. The implication of the analysis argued for the Klasies data is that transport was more likely in terms of processing motives. Items requiring processing were collected opportunistically in the context of regular feeding at places of procurement. We might even imagine that the collection of items requiring processing would only occur when foods *not* requiring processing were unavailable. If true, the transport of items requiring processing may well betray types of food stress that could vary seasonally. In any event, once collected, a potential food requiring processing would be carried to the place where the processing could be readily performed. In the case of the animal parts remaining at sites of ravaged carcasses, at least one processing prerequisite would be usable hammers and probably anvils. Another aid to processing might be water, in which to soak badly desiccated encasing skin so that cutting tools could be used for exposing the bones prior to breaking them open for marrow.

The consumption of naturally dried tissue adhering to already ravaged bones could be facilitated by having effective scraper–cutting tools for detaching the stringy morsels, or water in which the dry meat could be both washed and reconstituted so that remnants could be more readily removed. If water was the first major tether to hominid movement, fire was almost certainly the second and perhaps more important tether. After the use of fire, many other processing options opened up. In its initial technological role, fire was most likely used to secure sleeping places and to provide warmth. It provided light and heat, and tended to discourage nocturnal predators such as the leopard and hyaena.

Early regular use of fires was almost certainly dependent upon maintaining a fire at the sleeping site. This could have been done by various fire-banking techniques, the most likely of which was simply to cover up glowing coals with ash so they could remain active for long periods of time. Rekindling was a simple matter of uncovering the embers and adding fuel. When moving to different sleeping locations, smouldering firebrands probably were transported. For cooking to be recognized as a processing alternative, it had to result from experimentation with fire. Once fire was understood, items judged appropriate to this form of processing would have had to be transported to the location of the fire—the sleeping location, since fire making was not a likely option open to diurnal feeding parties.

Data from Klasies River are most consistent with the view that parts processed by cooking were mainly anatomical units that could be transported as contained units—that is, in which the skin, after butchering, still functioned as an enclosing natural container. Cooking in its own skin served to preserve the juices inherent in fresh meat. Thus, without manufactured containers, humans found containers in the form of small animals that could be cooked complete (see descriptions of cooking this way in Australia in Binford 1983b:165–169), or parts could be butchered from large animals that preserved this natural container of skin. The rear leg, head, and neck were the most obvious units that could be cut off, still leaving most of the meat encased in skin. It is my guess that it is in the context of consuming food processed by cooking that regular sharing began as a common hominid social trait.

The presence of a water source in what was most likely primarily a sleeping area at Klasies River Mouth may well account for the high frequency of scavenged animal parts as well as parts returned for cooking. Other sleeping sites, not having available water, may well have a different assemblage of introduced anatomical parts, varying with the types of processing that could be carried out at the site.

Once the assumption of transport for sharing is challenged, we can readily recognize that if it is accepted as plausible that transport could be in terms of other goals or motives, then the methodological challenge is to reduce the potential ambiguity. In other words, we must develop ways of recognizing unambiguously (not passing judgment on what is most likely; see Binford 1983c) transport–processing phenomenon from transport–sharing. While I have already suggested some research directions relating to food-package sizes, other research directions can certainly be recognized and developed.

If we can reasonably imagine contexts of transport that do not imply sharing, can we imagine plausible hominid land-use patterns that do not require central-place-foraging forms of organized habitat use? I think the answer is a resounding *yes!*

It seems to me that there are several clues to the character of early hominid adaptation implied by the data from Klasies River Mouth. I have noted earlier that there is a surprising lack of flexibility indicated by the redundancy of the deposits laid down over relatively long periods of time. In marked contrast, I have tried to show how, among modern hunters and gatherers, differential positioning of the life functions of a group within the overall geographical range used by that group is done fairly commonly, and it changes the economic utility of places relative to the new position of the system within its range (see Binford 1982b, 1983a). This can be understood partially as changes in the transport costs that might characterize the use of places by hominids when they change their position in habitat.

What seems clear from the vast accumulation of deposits at the various caves and shelters at Klasies River Mouth is that, once the hominids began using this location, they used it tenaciously, until the sheltered areas were actually filled up with the debris from their prior use so that the place was no longer usable as a sheltered location. Certainly, throughout the period of use resulting in the accumulation at Cave 1, there does not appear to have been any change in the way the place was used. A land-use pattern characterized as *entrenched* and *routed* might be an appropriate way of imagining the mobility dynamics standing behind such an unvarying and tenacious use of specific places.

One can imagine a system in which there are certain stationary and relatively stable features of the environment that are of critical importance to a hominid group: (1) a water source that is also an ecologically important place to other species; (2) a protected sleeping place, such as a cave or rockshelter, generally away from the normal routes taken by nocturnal predators; and (3) a lithic source of raw materials for tools, or items usable directly. Between these *places* would be *zones,* or generalized settings, where a hominid could expect to find or encounter a variety of foods. The hominid can then be thought of as feeding among the places where there is appropriate sleeping space, safe drinking, or sources of materials for tools, and so forth. Some environments would offer differing degrees of point specificity to the places critical to a hominid's success. In settings where water sources were plentiful, though quite diffuse, their distribution might have had little influence on the routing of the hominids through their habitat. In other settings, however, the scarcity of water sources might have made them critical places, relative to which hominids were routed in their movements over extremely long periods of time.

The routed aspect of the system, as suggested here, is in recognition that we can imagine a time when hominids did not make shelters, and hence would not have been able to ensure adequate shelter at the same place where adequate water might be located. Similarly, they may not have had long-term planning standing behind their procurement of raw materials, which among modern men makes possible the continuous, coincident, and coterminus distribution of tools and tasks. Similarly, the relationships among the basic needs of life space, protected sleeping areas, resting areas, food, and water can be thought of as necessarily routing the movement of the hominids among the spots where their basic needs could be fulfilled by virtue of the particular spatial structure within their environment. If these spatially differentiated places did in fact provide different, critically important materials and conditions for the hominids, then the *routing* of the hominids among these places would be *entrenched*. Hominids would move among the places in almost unvarying patterns, by virtue of there being no way of altering the spatial structure of these critical conditions and materials to serve their own needs.

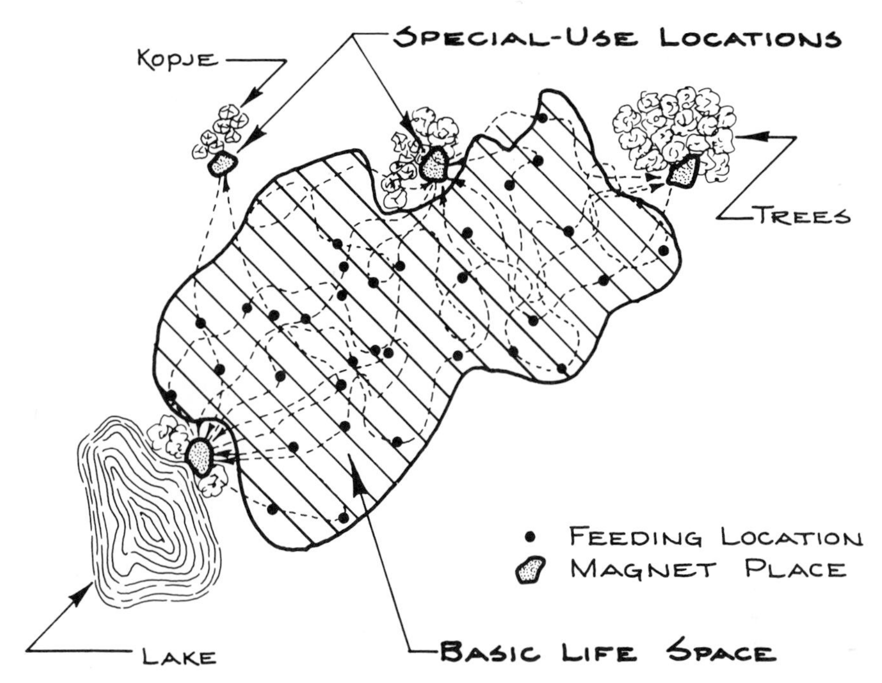

Figure 6.4 Routed feeding land-use model for early hominid site production.

The degree of entrenchment may also vary depending upon the frequency of places in home ranges offering roughly equivalent resources or conditions. Environmental variability would directly condition different home-range sizes and minor differences in the regularity with which places would be used. It would directly condition the combinations of life functions and activities that would be carried out at any one place. Although the model acknowledges that environmental variability can condition very different realized patterns of land use, the pattern in any given enviornmental setting would be expected to be relatively unvarying and tenacious, disruption of the pattern of use being most generally a response to changes in the environment.

The result of such a routed and entrenched adaptation is a very territorial pattern of land use. Hominids would occupy the same basic range with the same core area, and their movements would be routed among the same critically important places. Only changes in the environment or technological changes permitting them to construct their life space in their own interest would modify the repetitive, almost robotic use of space. I have modeled this type of system in Figure 6.4, in which the point-specific places providing critically important materials or resources are indicated as *special-use locations* (magnet places), and the basic life space is the range within which more diffuse resources, particularly foods, may be encountered as the hominid moves among the special-use locations. The sites are not places where hominids are constructing or modifying the environment in service of their vital needs, but instead are the point-specific locations in the local habitat where critically important resources or life conditions are available in nature. Because of the critical importance of these conditions, the use of such locations is regular, repetitive, and frequent. If the creature using these locations is capable both physically and motivationally of carrying things, then we can expect there to be an increasing concentration of transported items at such locations as a function of the regularity with which such places are used. Sites, under the conditions of this model of land use, are not "central places," positioned so the hominid group may operate out into the environment in terms of mobility and labor concerns; they are instead focal points in the environments of the hominids. In one very real sense, we can think of cultural evolution as generating a pattern of unrelenting intensification of facilities basic to hominid functions in increasingly complex coterminus arrangements. The abilities to construct shelters and to use fire made it possible for man to localize his sleeping activities adjacent to other critically important materials, such as water or food. With such intensification of functions, hominids became less routed among the places where nature had placed the elements basic to life functions. With the technological means now to construct features to meet vital needs, hominids could increasingly position their activities with respect to labor requirement. Positioning encouraged the provisioning of central places through the accumulation of materials from the environment rather than the simple placing of life functions passively within the habitat.

A major point is that the construction of increasingly portable life-space permitted greater mobility in hominid adaptations. This mobility permitted positioning strategies commonly associated with a central-place foraging strategy. The implication of portable life-space in a more general sense is that it permits more flexible and timely responses to ecological periodicities in both the production and distribution of foods upon which hominids depend. This type of ecological responsiveness is not easily argued for the early time ranges. Economically responsive, goal-directed behavior may well be unique to modern cultural man. Reading this behavior into the behavior standing behind the archaeological sites of ancient hominids could be a serious error.

Some Final Thoughts

I have emphasized the mobility tactics of a central-place foraging strategy, and suggested that it was the gradual obviating of certain environmental limitations on organized behavior that made possible the home-base pattern that we see in modern man. I am of the opinion that intelligence was a prerequisite to the technological solution to adaptive problems. The hominid had to recognize the problem, and then experiment with his knowledge to invent a solution. The selection for intelligence generally had to precede the use of the intelligence in complex problem-solving. Thus, I am suggesting that technological innovations and tool use were generally dependent upon high intelligence levels, and were therefore not the fundamental behavioral contexts in which selection for intelligence occurred. The arguments for selection of large brains and marked intelligence must precede, in general, the growth of technological aids in adaptation. The technical aids did, however, reduce the environmental limitation on forms of organized behavior, and many of the changes in such behavior that marked the history of the hominids after the appearance of tools will be understandable in terms of technologically modified relationships between the species and its environment, altering considerably the overall ecological role that hominids played in their environments and, in turn, the kinds and magnitude of selective pressures affecting them.

As was pointed out in the introduction, the consensus view organized against the aggressive-hunter model of early man was that of man as a "transitional" scavenger and opportunistic eater of baby pigs and birds' eggs. The transitional model was thought to apply to the human ancestor living at the Plio–Pleistocene boundary. I have suggested that the discovery of the Zinjanthropus floor forced a reappraisal. The floor was littered with the bones of large animals. The Plio–Pleistocene ancestor clearly seemed to have been taking large game. The inference that man was the predatory agent responsible for the bones found in association with his earliest tools forced the conclusion that our intelligence and our "humanness" had to have appeared prior to the time of the Olduvai levels. This was considered a given.

Ever since the dust settled down on the Olduvai excavations, the trend has been to push the processes of humanization further and further into the past. Going hand in hand with this trend has been an awakened understanding of the possible role of sexual selection, and a heretofore unsuspected role for social behavior in natural selection. Interesting models of evolutionary transformation have been constructed to accommodate a historical pattern that assumed a hunting strategy, with animal parts transported to a central place, presumably for the purpose of sharing with others. Delightful examples of this type of model building are Helen Fisher's fascinating book, *The Sex Contract* (1983), and the "gathering model" of Nancy Tanner (1981). As more such models are built, the archaeologist working essentially without any intellectually independent methodology for inference will most certainly accommodate his findings to the beliefs of his choice. I have presented a methodological approach that leads us to conclude that the sites that were taken as self-evident testimony to hunting, home bases, and transport for sharing, no longer appear so self-evident in their meanings. In fact, most of the Olduvai sites now appear to represent midday-rest locations where scavenged morsels were carried by hominids for processing and the recovery of only marginal food tidbits. The tiny units of food indicated hardly qualify as a realistic package to be shared with others in the sense that hunting models of sharing generally assume.

The picture at Klasies River Mouth of a hominid living very close to the era of behaviorally modern man, but behaving in a manner that does not appear to be vastly different from the behavior of hominids living on the Plio–Pleistocene boundary, is surprising. If nothing else, it brings into question the timing within current evolutionary models designed to account for our humanity. As more research is designed and carried out to learn about the past, we may find that modifications in our evolutionary models will also be in order. The use of experience to evaluate our ideas is the basic program of science. Archaeology must, through the development of robust inferential methods, make it possible for the past to "talk back" to our ideas regarding its character. Only when the past can be given a voice to argue with our ideas of its character will we truly begin to unravel the fascinating processes of our own becoming.

References

Adam, K.

1951 Der Waldelefant von Lehringen, eine Jagdbeute des diluvialen Menschen. *Quartär* 5:75–92.

Altuna, J.

1971 Los mamíferos del yacimiento prehistórico de Morin (Santander). In *Cueva Morin: Excavaciones 1966–1968,* edited by J. G. Echegaray, and L. G. Freeman. Publicaciones del patronato de las cuevas prehistóricas del la provincia de Santander. Pp. 369–400.

Ayeni, J. S. O.

1975 Utilization of waterholes in Tsavo National Park (east). *East African Wildlife Journal* 13:305–323.

Bartholomew, G. A., and J. B. Birdsell

1953 Ecology and the protohominids. *American Anthropologist* 55(4):481–496.

Beaumont, P. B.

1980 On the age of Border Cave hominids 1–5 *Paleontologia Africana* 23:21–33.

Beaumont, P. B., and J. C. Vogel

1972 On a new radiocarbon chronology for Africa south of the equator. *African Studies* 31:65–80, 155–182.

Beaumont, P. B., H. R. de Villiers, and J. C. Vogel

1978 Modern man in sub-Saharan Africa prior to 49,000 years B.P.: A review and evaluation with particular reference to Border Cave. *South African Journal of Science* 74:409–419.

Behrensmeyer, A. K., and D. E. Dechant Boaz.

1980 The recent bones of Amboseli National Park, Kenya, in relation to East African Paleoecology. In *Fossils in the making: vertebrate taphonomy and paleoecology,* edited by A. K. Behrensmeyer and A. P. Hill. Chicago & London:University of Chicago Press. Pp. 72–93.

Bertram, B.
1978 *Pride of lions.* New York:Scribner.

Binford, L. R.
1962 Archaeology as Anthropology. *American Antiquity* 28(2):217–225.
1963 An analysis of cremations from three Michigan sites. *Wisconsin Archeologist* 44:98–110.
1968 Archaeological theory and method. In *New perspectives in archaeology,* edited by S. R. Binford and L. R. Binford. Chicago:Aldine. Pp. 1–3.
1977a General introduction. In *For theory building in archaeology,* edited by L. R. Binford. New York:Academic Press. Pp. 1–10.
1977b Olorgesailie deserves more than the usual book review. *Journal of Anthropological Research* 33(4):493–502.
1978 *Nunamiut ethnoarchaeology.* New York:Academic Press.
1980a Review of "Reindeer; and Caribou Hunters" by A. E. Spiess. *American Anthropologist* 82(3):628–631.
1980b Willow smoke and dog's tails: hunter–gatherer settlement systems and archaeological site formation. *American Antiquity* 45(1):1–17.
1981 *Bones:ancient men and modern myths.* New York:Academic Press.
1982a Comments on R. White's 'Middle/Upper paleolithic transition.' *Current Anthropology* 23(2):177–181.
1982b The Archaeology of place. *Journal of Anthropological Archaeology* 1(1):5–31.
1983a *Working at archaeology.* New York:Academic Press.
1983b *In pursuit of the past.* London:Thames & Hudson.
1983c Reply to L. Freeman. *Current Anthropology* 24(3):372–376.
1984 Bones of contention—a reply to Glynn Isaac, *American Antiquity* 49(1). *In press.*

Binford, L. R., and J. Bertram
1977 Bone frequencies and attritional processes. In *For theory building in archaeology,* edited by L. R. Binford. New York:Academic Press. Pp. 77–153.

Binford, L. R., and J. F. O'Connell
In press An Alyawara day: the stone quarry. *Journal of Anthropological Archaeology.*

Binford L. R., and L. Todd
1982 On arguments for the 'butchering' of giant Geladas. *Current Anthropology* 23(1):108–110.

Bonifay, M. F.
1969 Les grands mammifères découverts sur le sol de la cabane Achéuleenne du Lazaret. In *Une cabane Acheuléenne dans la grotte du Lazaret (Nice),* edited by H. De Lumley. *Mémoires de la société préhistorique Française,* Tome 7. Paris:C.N.R.S. Pp. 59

Bordes, F.
1961 Mousterian cultures in France. *Science* 134(3482):803–810.

Brain, C. K.
1969 Faunal remains from the Wilton large rock shelter. In *Re-excavation and description of the Wilton type-site, Albany district, Eastern Cape* by J. Deacon. M. A. thesis, University of Cape Town.
1974 Some suggested procedures in the analysis of bone accumulations from Southern African Quaternary sites. *Annals of the Transvaal Museum* 29:1–5.
1981 *The hunters or the hunted? an introduction to African cave taphonomy.* Chicago/London:University of Chicago Press.

Bunn, H.
1981 Archaeological evidence for meat-eating: by Plio-Pleistocene hominids from Koobi Fora and Olduvai Gorge. *Nature* 291:574–577.
1982 Animal bones and archaeological inference. *Science* 215:494–495.
Bunn, H., J. W. K. Harris, G. Isaac, A. Kaufulu, E. Kroll, K. Schick, N. Toth, and A. K. Behrensmeyer
1980 FxJ50: An Early Pleistocene site in Northern Kenya. *World Archaeology* 12(2):109–136.
Butzer, K. W.
1973 Geology of Nelson Bay Cave, Robberg, South Africa. *South African Archaeological Bulletin* 28:97–110.
1978 Sediment stratigraphy of the Middle Stone Age sequence at Klasies River Mouth, Tsitsikama Coast, South Africa. *South African Archaeological Bulletin* 33:141–151.
1979 Comment on "Implications of Border Cave skeletal remains for later Pleistocene human evolution." *Current Anthropology* 20:28.
1982 Geomorphology and sediment stratigraphy. In *The Middle Stone Age at Klasies River Mouth in South Africa,* R. Singer and J. Wymer. Chicago and London, The University of Chicago Press. Pp. 33–42.
Butzer, K. W., P. B. Beaumont, and J. C. Vogel
1978 Lithostratigraphy of Border Cave, KwaZulu, South Africa: a Middle Stone Age sequence beginning ca. 195,000 B.P. *Journal of Archaeological Science* 5:317–341.
Butzer, K. W., and D. M. Helgren
1972 Late Cenozoic evolution of the Cape coast between Knysna and Cape St. Francis, South Africa. *Quarternary Research* 2:143–169.
Campbell, B. G.
1966 *Human Evolution.* Chicago/New York:Aldine–Atherton Press.
Chaplin, R. E.
1971 *The study of animal bones from archaeological sites.* New York/London:Seminar Press.
Clark, J. D.
1959 *The prehistory of Southern Africa.* London:Pelican.
1965 Pleistocene culture and living sites. In *The origin of man—a symposium sponsored by the Wenner–Gren Foundation for Anthropological Research, Inc.,* edited by P. L. DeVore. Chicago:Distributed through *Current Anthropology.* Pp. 108–123.
1980 Early human occupation of African Savanna Environments. In *Human ecology in savanna environments,* edited by D. R. Harris. London, New York:Academic Press. Pp. 41–71.
Dart, R.
1949 The predatory implemental technique of *Australopithecus. American Journal of Physical Anthropology* 7(1):1–39.
Darwin, C.
1839 *The Voyage of the Beagle.* Garden City, New York:Doubleday/Anchor edition, 1962.
Day, J., W. R. Siegfried, G. N. Louw, and M. L. Jarman (editors)
1979 Fynbos ecology: a preliminary synthesis. Pretoria. *South African National Scientific Programmes Report* No. 40.

Deacon, H., and J. Deacon
1980 The hafting, function and distribution of small convex scrapers with an example from Boomplaas Cave. *South African Archaeological Bulletin* 35(131): 31–37.
1981 Some novel news on Neanderthals. *The South African Archaeological Society Newsletter* 4(2):6–8.
De Lumley, H.
1969 A Paleolithic camp at Nice. *Scientific American* 220(5):42–50.
1975 Cultural evolution in France in its paleoecological setting during the Middle Pleistocene. In *After the Australopithecines,* edited by K. W. Butzer and G. Ll. Isaac. The Hague, Paris, & Chicago:Mouton Publishers. Pp. 745–808.
De Lumley, H. (editor)
1969 *Une cabane Acheuléenne dans la grotte du Lazaret (Nice) Mémoires de la Société Préhistorique Française,* Tome 7. Paris: C.N.R.S.
De Lumley, H., and Y. Boone
1976 Les structures d'habitat au Paléolithique inférieur. In *La prehistorire française,* edited by H. de Lumley. Paris:Éditions C.N.R.S. Pp. 625–643.
Etkin, W.
1954 Social behavior and the evolution of man's mental faculties. *The American Naturalist* 88(840):129–142.
Fisher, H. E.
1983 *The sex contract.* New York:Morrow/Quill.
Freeman, L. G.
1983 More on the Mousterian: flaked bone from Cueva Morin. *Current Anthropology* 24(3):366–372.
Gould, S. J.
1981 *The mismeasure of man.* New York & London:Norton.
Guilday, J. E., P. Parmalee, and D. Tanner
1962 Aboriginal butchering techniques at the Eschelman site (36 LA 12) Lancaster County. *Pennsylvania Archaeologist* 32:59–83.
Hayden, B.
1979 *Palaeolithic reflections.* New Jersey:U.S.A. Humanities Press.
Haynes, G.
1982 Utilization and skeletal disturbances of North American prey carcasses. *Arctic* 35(2):266–281.
1983 Frequencies of spiral and green bone fractures on ungulate limb bones in modern surface assemblages. *American Antiquity* 48(1):102–114.
Hill, A. P.
1975 *Taphonomy of contemporary and Late Cenezoic East African vertebrates.* Unpublished Ph.D. dissertation, University of London.
1979 Butchery and natural disarticulation: an investigatory technique. *American Antiquity* 44(4):739–744.
Howell, F. C., and the Editors of Time–Life Books
1965 *Early man.* New York:Time–Life Books.
Inskeep, R. R.
1978 *The Peopling of Southern Africa.* Capetown: David Philip.
Isaac, G. Ll.
1971 The diet of early man: aspects of archaeological evidence from Lower and Middle Pleistocene sites in Africa. *World Archaeology* 2:278–298.
1975a Early hominids in action: a commentary on the contribution of archaeology to

understanding tbe fossil record in East Africa. *Yearbook of Physical Anthropology* 19, edited by J. Buettner–Janusch. Pp. 178–191.

1975b Early stone tools—an adaptive threshold? In *Problems in economic and social archaeology,* edited by F. de G. Sieveking, I. H. Longworth, and K. E. Wilson. London:Duckworth. Pp. 39–47.

1976 Stages of cultural elaboration in the Pleistocene: possible archaeological indications of the development of language capabilities. In *Origins and evolution of language and speech,* edited by S. R. Harnad, H. D. Steklis, and J. Lancaster. Annals of the New York Academy of Sciences 280:275–288.

1977 *Olorgesailie: archaeological studies of a Middle Pleistocene lake basin in Kenya.* Chicago/London:University of Chicago Press.

1978a The food-sharing behavior of protohuman hominids. *Scientific American* 238(4):90–106.

1978b Food sharing and human evolution: archaeological evidence from the Plio-Pleistocene of East Africa. *Journal of Anthropological Research* 34(3): 311–325.

1983a Review of "Bones: Ancient men and modern myths." *American Antiquity* 48(2):416–419.

1983b Bones in contention: competing explanations for the juxtaposition of Early Pleistocene artifacts and faunal remains. In *Animals and archaeology: Hunters and their prey,* edited by J. Clutton-Brock and C. Grigson. Oxford:British Archaeological Reports 163:1–19.

Isaac, G. and D. Crader

1981 To what extent were early hominids carnivorous? An archaeological perspective. In *Omnivorous primates: Gathering and hunting in human evolution,* edited by R. S. O. Harding and G. Teleki. New York:Columbia University Press. Pp. 37–103.

Jarvis, J. U. M.

1972 Zoogeography. In *Fynbos ecology: a preliminary synthesis,* edited by J. Day *et al.* Pretoria: South African national scientific programmes report No. 40. Pp. 82–87.

Jelinek, A. J.

1982 The Tabun Cave and Paleolithic man in the Levant. *Science* 216(4553): 1369–1375.

Kehoe, T. F.

1967 The Boarding School Bison Drive site. *Plains Anthropologist Memoir* No. 4.

Kingdon, J.

1982 *East African mammals (Vol. III, Part C [Bovids]).* London–New York:Academic Press.

Klein, R. G.

1970 Problems in the study of the Middle Stone Age of South Africa. *South African Archaeological Bulletin* 25(99 & 100):127–135.

1972 The late quaternary mammalian fauna of Nelson Bay Cave (Cape Province, South Africa); its implications for megafaunal extinctions and environmental and cultural change. *Quaternary Research* 2(2):135–142.

1974 Environment and subsistence of prehistoric man in the Southern Cape Province, South Africa. *World Archaeology* 5(3):249–284.

1975a Ecology of Stone Age man at the southern tip of Africa. *Archaeology* 28:238–247.

1975b Middle Stone Age: man–animal relationships in southern Africa: evidence from Die Kelders and Klasies River Mouth. *Science* 190:265–267.
1976 The mammalian fauna of the Klasies River Mouth sites, Southern Cape Province, South Africa. *South African Archaeological Bulletin* 31(123 & 124):75–98.
1977a The ecology of early man in South Africa. *Science* 197:115–126.
1977b The mammalian fauna from the Middle and Later Stone Age (Later Pleistocene) levels of Border Cave, Natal Province, South Africa. *South African Archaeological Bulletin* 32:14–27.
1978a Preliminary analysis of the mammalian fauna from the Redcliff Stone Age cave site, Rhodesia. *Occasional Papers of the National Museums and Monuments of Rhodesia Series A. Human Sciences,* 4(2):74–80.
1978b A preliminary report on the larger mammals from the Boomplaas Stone Age cave site, Cango Valley, Oudtshoorn district, South Africa. *South African Archaeological Bulletin* 33:66–75.
1978c Stone Age predation on large African bovids. *Journal of Archaeological Science* 5:195–217.
1979 Stone Age exploitation of animals in southern Africa *American Scientist* 67:151–160.
1980 The interpretation of mammalian faunas from Stone Age archaeological sites, with special reference to sites in the Southern Cape Province, South Africa. In *Fossils in the making: vertebrate taphonomy and paleoecology,* edited by A. K. Behrensmeyer and A. P. Hill. Chicago & London:University of Chicago Press. Pp. 223–246.
1981 Stone Age predation of small African bovids. *South African Archaeological Bulletin* 36:55–65.
1982 Age (mortality) profiles as a means of distinguishing hunted species from scavenged ones in Stone Age archaeological sites. *Paleobiology* 8(2):151–158.

Kruger, F. J.
1979 Plantecology. In *Fynbos ecology: a preliminary synthesis,* edited by J. Day *et al.* Pretoria: South African National Scientific Programmes Report No. 40. Pp. 88–126.

Kuhn, T.
1970 *The structure of scientific revolutions* (2nd edition) Chicago:University of Chicago Press.

Leakey, L. S. B.
1960 The origin of the genus *Homo. Evolution after Darwin (Vol. II: The Evolution of Man,* edited by Sol Tax. Chicago:University of Chicago Press. Pp. 17–32.
1965 Facts instead of dogmas on man's origin. In *The origin of man.* Transcript of a symposium sponsored by the Wenner–Gren Foundation for Anthropological Research. Edited by Paul L. Devore. Pp. 3–17.

Leakey, M. D.
1971 *Olduvai Gorge (Vol. 3 Excavations in Beds I and II 1960–1963).* Cambridge: Cambridge University Press.

Leakey, R. E.
1973 Skull 1470: discovery in Kenya of the earliest suggestion of the genus *Homo*—nearly three million years old—compels a rethinking of mankind's pedigree. *National Geographic* 143(6):819–829.
1981 *The making of mankind.* New York:Dutton.

Lent, P.
1974 Mother–infant relationships in ungulates. In *Behavior of ungulates and its*

relation to management, edited by V. Geist and F. Walther (New Series Number 24) Morges: International Union for the Conservation of Nature and Natural Resources. Pp. 14–55.

Lewin, R.
1981 AAAS briefings: Ancient cut marks reveal work of prehuman hands. *Science* 211:372–373.

Lovejoy, C. O.
1981 The origin of man. *Science* 211(4480):341–350.

McBurney, C. B. M.
1967 *The Haua Fteah (Cyreneica) and the Stone Age of the South-East Mediterranean.* Cambridge:Cambridge University Press.

Maguire, J. M. and D. Pemberton
1980 The Makapansgat limeworks grey breccia: hominids, hyaenas, hystricids or hillwash? *Paleontologia Africana* 23:75–98.

Marshall, L.
1976 *The !Kung of Nyae Nyae.* Cambridge:Harvard University Press.

Mech, L. D.
1966 *The wolves of Isle Royale. Fauna of the National Parks of the United States,* Series 7. Washington D. C.:U.S. Government Printing Office.

Mellars, P.
1982 On the Middle/Upper Palaeolithic transition: A reply to White. *Current Anthropology* 23(2):238–240.

Moss, C.
1982 *Portraits in the wild* (2nd edition). Chicago:University of Chicago Press.

Parkington, J. F.
1972 Seasonal mobility in the Late Stone Age. *African Studies* 31:223–243.
1976 Coastal settlement between the mouths of the Berg and Olifants rivers, Cape Province. *South Africa Archaeological Bulletin* 31:127–140.
1980 Time and place: Some observations on spatial and temporal patterning in the Later Stone Age sequence in southern Africa. *South African Archaeological Bulletin* 35(132):73–83.
1981 The effects of environmental change on the scheduling of visits to the Elands Bay Cave, Cape Province, South Africa. In *Pattern of the Past: Studies in Honor of David Clarke,* edited by I. Hodder, G. Ll. Isaac, and N. Hammond. Cambridge: Cambridge University Press. Pp. 341–359.

Parkington, J., and C. Poggenpoel
1971 Excavations at De Hangen, 1968. *South African Archaeological Bulletin* 26(101–102):3–36.

Perkins, D., and P. Daly
1968 A hunters' village in Neolithic Turkey. *Scientific American* 219(5):97–106.

Potts, R. B.
1982 *Lower Pleistocene site formation and hominid activities at Olduvai Gorge, Tanzania.* Doctoral thesis, Department of Anthropology, Harvard University, Cambridge, MA.

Potts, R., and P. Shipman
1981 Cutmarks made by stone tools on bones from Olduvai Gorge, Tanzania. *Nature* 291:577–580.

Read, C.
1920 *The origin of man and his superstitions.* Cambridge:Cambridge University Press.

Richardson, P. R. K.
1980a *The natural removal of ungulate carcasses, and the adaptive features of the scavengers involved.* Unpublished masters thesis, Faculty of Science, University of Pretoria, South Africa.
1980b Carnivore damage on antelope bones and its archaeological implications. *Paleontologia Africana* 23:109–125.
Rightmire, G. P.
1976 Relationships of Middle and Upper Pleistocene hominids from sub-Saharan Africa. *Nature* 260:238–240.
1978 Human skeletal remains from the Southern Cape Province and their bearing on the Stone Age prehistory of South Africa. *Quaternary Research* 9:219–230.
1979 Implications of Border Cave skeletal remains for Later Pleistocene human evolution. *Current Anthropology* 20(1):23–35.
Sampson, C. G.
1974 *The Stone Age archaeology of Southern Africa.* New York and London: Academic Press.
Schalke, H. J.
1973 The upper quaternary of the Cape Flats areas (Cape Province, South Africa). *Scripta Geoligica* 15:1–57.
Schaller, G. B.
1972a Are you running with me, hominid? *Natural History* 81:60–69.
1972b *The Serengeti lion.* Chicago:University of Chicago Press.
Schaller, G. B., and G. Lowther
1969 The relevance of carnivore behavior to the study of early hominids. *Southwestern Journal of Anthropology* 25(4):307–341.
Schiffer, M.
1972 Archaeological context and systematic context. *American Antiquity* 37:156–165.
Shackleton, N. J.
1975 The stratigraphic record of deep-sea cores and its implications for the assessment of glacials, interglacials stadials, and interstadials in the Mid-Pleistocene. In *After the australopithecines,* edited by K. W. Butzer and G. Ll. Isaac. The Hague & Paris:Mouton Publishers. Pp. 1–24.
1982 Stratigraphy and chronology of the KRM deposits: oxygen isotope evidence. In *The Middle Stone Age at Klasies River Mouth in South Africa,* edited by Ronald Singer and John Wymer, Chicago:University of Chicago Press. Pp. 194–199.
Shackleton, N. J., and N. D. Opdyke
1973 Oxygen-isotope and palaeomagnetic stratigraphy of equatorial Pacific core V28–238: oxygen isotope temperatures and ice volumes on a 10^5 year and 10^6 year scale. *Quaternary Research* 3:39–55.
Shipman, P., W. Bosler, and K. L. Davis
1981 Butchering of giant Geladas at an Acheulian site. *Current Anthropology* 22(3):257–264.
1982 Reply to Binford and Todd. *Current Anthropology* 23(1):110–111.
Shipman, P., and J. Phillips–Conroy
1977 Hominid tool-making versus carnivore scavenging. *American Journal of Physical Anthropology* 46(1):77–86.
Sinclair, A. R. E.
1977 *The African buffalo: a study of resource limitation of populations.* Chicago & London:The University of Chicago Press.

Sinclair, A. R. E., and M. Norton–Griffiths (editors)
1979 *Serengeti: Dynamics of an ecosystem.* Chicago:University of Chicago Press.
Singer, R., and P. Smith
1969 Some human remains associated with the Middle Stone Age deposits at Klasies River, South Africa. *American Journal of Physical Anthropology* 31:256.
Singer, R., and J. Wymer
1982 *The Middle Stone Age at Klasies River Mouth in South Africa.* Chicago/London:University of Chicago Press.
Skinner, J. D., S. Davis, and G. Ilani
1980 Bone collecting by striped hyaenas, *Hyaena hyaena* in Israel. *Paleontologia Africana* 23:99–104.
Spencer, B., and F. J. Gillen
1912 *Across Australia* (Vol. II). London:Macmillan.
Speth, J. D.
1983 *Bison kills and bone counts: decision making by ancient hunters.* Chicago & London:University of Chicago Press.
Spinage, C. A.
1982 *A territorial antelope: the Uganda waterbuck.* London & New York:Academic Press.
Steyn, H. P.
1971 Aspects of the economic life of some nomadic Nharo Bushman Groups. *Annals of the South African Museum* 56(6):275–322.
Stow, G.
1905 *The native races of South Africa.* London:Swan Sonnenschein.
Tankard, A. J.
1976 The stratigraphy of a coastal cave and its palaeoclimatic significance. *Palaeoecology of Africa* 9:151–159.
Tankard, A. J., and Rogers, J.
1978 Late Cenozoic palaeoenvironments on the west coast of southern Africa. *Journal of Biogeography* 5:319–337.
Tanner, N. M.
1981 *On becoming human.* Cambridge:Cambridge University Press.
Thackeray, F.
1979 An analysis of faunal remains from archaeological sites in southern South West Africa (Namibia). *South African Archaeological Bulletin* 34(129):18–33.
van Lawick–Goodall, J., and H. van Lawick
1970 *Innocent killers.* Boston:Houghton-Mifflin Co.
van Zinderen Bakker, E. M.
1967 Upper Pleistocene and Holocens stratigraphy and ecology on the basis of vegetation changes in sub-Saharan Africa. In *Background to evolution in Africa,* edited by W. Bishop and J. D. Clark. Chicago:University of Chicago Press. Pp. 125–147.
1976 The evolution of late-Quaternary paleoclimates of southern Africa. *Palaeoecology of Africa* 9:160–202.
Vogel, J. C., and Beaumont, P. B.
1972 Revised radiocarbon chronology for the Stone Age in southern Africa. *Nature* 237:50–51.
Voigt, E. A.
1973a Stone age molluscan utilization at Klasies River Mouth caves. *South African Journal of Science* 69:306–309.

1973b Klasies River Mouth: an exercise in shell analysis. *Bulletin of the Transvaal Museum.* 14:14–15.

Volman, T. P.

1978 Early archaeological evidence for shellfish collecting. *Science* 210:911–913.

1981 *The Middle Stone Age in the Southern Cape.* Ph.D. dissertation submitted to the Faculty of the Department of Anthropology, University of Chicago, Chicago, Il.

Vrba, E. S.

1976 *The fossil bovidae of Sterkfontein, Swarthkrans, and Kromdraai, Transvaal Museum Memoir* No. 21. The Transvaal Museum, Pretoria.

Walton, J.

1951 Occupied rock-shelters in Basutoland. *South African Bulletin* 4(21):9–13.

Washburn, S. L.

1959 Speculations on tbe interrelations of the history of tools and biological evolution. In *The evolution of man's capacity for culture,* edited by J. N. Spuhler Detroit:Wayne State University Press. Pp. 21–31.

Washburn, S. L., and F. C. Howell

1960 Human evolution and culture. *Evolution after Darwin (Vol. II. The evolution of man),* edited by Sol Tax. Chicago:University of Chicago Press. Pp. 33–56.

Washburn, S. L., and C. S. Lancaster

1968 The evolution of hunting. In *Man the hunter,* edited by R. B. Lee and I. DeVore. Chicago:Aldine. Pp. 293–303.

White, R.

1982 Rethinking the Middle/Upper Paleolithic transition. *Current Anthropology* Vol. 23(2)Pp. 169–176.

White, T. E.

1954 Observations on the butchering technique of some aboriginal peoples, Nos. 3, 4, 5, & 6. *American Antiquity* 19(3):254–264.

Wymer, J., and R. Singer

1972 Middle Stone Age settlements on the Tzitzikamma coast, Eastern Cape Province, South Africa. In *Man, settlement and urbanism,* edited by P. Ucko, R. Tringham, and G. Dimbleby. London:Duckworth. Pp. 207–210.

Yellen, J.

1977 *Archaeological approaches to the present: models for reconstructing the past.* New York:Academic Press.

Young, J. Z.

1981 Some tentative conclusions. In *The emergence of man,* edited by J. Z. Young, W. M. Jope, and K. P. Oakley. London:The Royal Society and the British Academy. Pp. 213–216.

Index

I

J

K

L

M

Y

Z

STUDIES IN ARCHAEOLOGY

Consulting Editor: Stuart Struever

Department of Anthropology
Northwestern University
Evanston, Illinois

Charles R. McGimsey III. **Public Archeology**

Lewis R. Binford. **An Archaeological Perspective**

Joseph W. Michels. **Dating Methods in Archaeology**

C. Garth Sampson. **The Stone Age Archaeology of Southern Africa**

Fred T. Plog. **The Study of Prehistoric Change**

Patty Jo Watson (Ed.). **Archaeology of the Mammoth Cave Area**

George C. Frison (Ed.). **The Casper Site: A Hell Gap Bison Kill on the High Plains**

W. Raymond Wood and R. Bruce McMillan (Eds.). **Prehistoric Man and His Environments: A Case Study in the Ozark Highland**

Kent V. Flannery (Ed.). **The Early Mesoamerican Village**

Charles E. Cleland (Ed.). **Cultural Change and Continuity: Essays in Honor of James Bennett Griffin**

Michael B. Schiffer. **Behavioral Archeology**

Fred Wendorf and Romuald Schild. **Prehistory of the Nile Valley**

Michael A. Jochim. **Hunter-Gatherer Subsistence and Settlement: A Predictive Model**

Stanley South. **Method and Theory in Historical Archeology**

Timothy K. Earle and Jonathon E. Ericson (Eds.). **Exchange System in Prehistory**

Stanley South (Ed.). **Research Strategies in Historical Archeology**

John E. Yellen. **Archaeological Approaches to the Present: Models for Reconstructing the Past**

Lewis R. Binford (Ed.). **For Theory Building in Archaeology: Essays on Faunal Remains, Aquatic Resources, Spatial Analysis, and Systemic Modeling**

James N. Hill and Joel Gunn (Eds.). **The Individual in Prehistory: Studies of Variability in Style in Prehistoric Technologies**

Michael B. Schiffer and George J. Gumerman (Eds.). **Conservation Archaeology: A Guide for Cultural Resource Management Studies**

Thomas F. King, Patricia Parker Hickman, and Gary Berg. **Anthropology in Historic Preservation: Caring for Culture's Clutter**

Richard E. Blanton. **Monte Albán: Settlement Patterns at the Ancient Zapotec Capital**

R. E. Taylor and Clement W. Meighan. **Chronologies in New World Archaeology**

Bruce D. Smith. **Prehistoric Patterns of Human Behavior: A Case Study in the Mississippi Valley**

Barbara L. Stark and Barbara Voorhies (Eds.). **Prehistoric Coastal Adaptations: The Economy and Ecology of Maritime Middle America**

Charles L. Redman, Mary Jane Berman, Edward V. Curtin, William T. Langhorne, Nina M. Versaggi, and Jeffery C. Wanser (Eds.). **Social Archeology: Beyond Subsistence and Dating**

Bruce D. Smith (Ed.). **Mississippian Settlement Patterns**

Lewis R. Binford. **Nunamiut Ethnoarchaeology**

J. Barto Arnold III and Robert Weddle. **The Nautical Archeology of Padre Island: The Spanish Shipwrecks of 1554**

Sarunas Milisauskas. **European Prehistory**

Brian Hayden (Ed.). **Lithic Use-Wear Analysis**

William T. Sanders, Jeffrey R. Parsons, and Robert S. Santley. **The Basin of Mexico: Ecological Processes in the Evolution of a Civilization**

David L. Clarke. **Analytical Archaeologist: Collected Papers of David L. Clarke. Edited and Introduced by His Colleagues**

Arthur E. Spiess. **Reindeer and Caribou Hunters: An Archaeological Study**

Elizabeth S. Wing and Antoinette B. Brown. **Paleonutrition: Method and Theory in Prehistoric Foodways**

John W. Rick. **Prehistoric Hunters of the High Andes**

Timothy K. Earle and Andrew L. Christenson (Eds.). **Modeling Change in Prehistoric Economics**

Thomas F. Lynch (Ed.). **Guitarrero Cave: Early Man in the Andes**

Fred Wendorf and Romuald Schild. **Prehistory of the Eastern Sahara**

Henri Laville, Jean-Philippe Rigaud, and James Sackett. **Rock Shelters of the Perigord: Stratigraphy and Archaeological Succession**

Duane C. Anderson and Holmes A. Semken, Jr. (Eds.). **The Cherokee Excavations: Holocene Ecology and Human Adaptations in Northwestern Iowa**

Anna Curtenius Roosevelt. **Parmana: Prehistoric Maize and Manioc Subsistence along the Amazon and Orinoco**

Fekri A. Hassan. **Demographic Archaeology**

G. Barker. **Landscape and Society: Prehistoric Central Italy**

Lewis R. Binford. **Bones: Ancient Men and Modern Myths**

Richard A. Gould and Michael B. Schiffer (Eds.). **Modern Material Culture: The Archaeology of Us**

Muriel Porter Weaver. **The Aztecs, Maya, and Their Predecessors: Archaeology of Mesoamerica, 2nd edition**

Arthur S. Keene. **Prehistoric Foraging in a Temperate Forest: A Linear Programming Model**

Ross H. Cordy. **A Study of Prehistoric Social Change: The Development of Complex Societies in the Hawaiian Islands**

C. Melvin Aikens and Takayasu Higuchi. **Prehistory of Japan**

Kent V. Flannery (Ed.). **Maya Subsistence: Studies in Memory of Dennis E. Puleston**

Dean R. Snow (Ed.). **Foundations of Northeast Archaeology**

Charles S. Spencer. **The Cuicatlán Cañada and Monte Albán: A Study of Primary State Formation**

Steadman Upham. **Polities and Power: An Economic and Political History of the Western Pueblo**

Carol Kramer. **Village Ethnoarchaeology: Rural Iran in Archaeological Perspective**

Michael J. O'Brien, Robert E. Warren, and Dennis E. Lewarch (Eds.). **The Cannon Reservoir Human Ecology Project: An Archaeological Study of Cultural Adaptations in the Southern Prairie Peninsula**

Jonathon E. Ericson and Timothy K. Earle (Eds.). **Contexts for Prehistoric Exchange**

Merrilee H. Salmon. **Philosophy and Archaeology**

Vincas P. Steponaitis. **Ceramics, Chronology, and Community Patterns: An Archaeological Study at Moundville**

George C. Frison and Dennis J. Stanford. **The Agate Basin Site: A Record of the Paleoindian Occupation of the Northwestern High Plains**

James A. Moore and Arthur S. Keene (Eds.). **Archaeological Hammers and Theories**

Lewis R. Binford. **Working at Archaeology**

William J. Folan, Ellen R. Kintz, and Laraine A. Fletcher. **Coba: A Classic Maya Metropolis**

David A. Freidel and Jeremy A. Sabloff. **Cozumel: Late Maya Settlement Patterns**

John M. O'Shea. **Mortuary Variability: An Archaeological Investigation**

Lewis R. Binford. **Faunal Remains From Klasies River Mouth**

in preparation

Robert I. Gilbert, Jr. and James H. Mielke (Eds.). **The Analysis of Prehistoric Diets**

John Hyslop. **The Inka Road System**

Christopher Carr (Ed.). **The Analysis of Archaeological Data Structures**